A Concise Manual
of Pathogenic
Microbiology

A Concise Manual of Pathogenic Microbiology

Saroj K. Mishra, Ph.D.
Dipti Agrawal, M.D.

WILEY-BLACKWELL

A John Wiley & Sons, Inc., Publication

Library of Congress Cataloging-in-Publication Data:
Mishra, Saroj (Saroj K.)
   A concise manual of pathogenic microbiology / Saroj K. Mishra, Dipti Agrawal.
      p. cm.
   Includes index.
   ISBN 978-1-118-30119-7 (hardback)
   1. Medical microbiology–Handbooks, manuals, etc.   2. Diagnostic microbiology–Handbooks,
manuals, etc.   I. Agrawal, Dipti.   II. Title.
   QR46.M597 2012
   616.9'041–dc23

                                                                              2012015252

Printed in the United States of America

10 9 8 7 6 5 4 3 2 1

Dedicated to the students of pathogenic microbiology

Contents

Preface

Pathogenic microbiology is a highly developed branch of microbiology. The past few decades have witnessed a quantum leap in our understanding of the molecular aspects of microorganisms and the host–microbe interactions. But the need for basic information on which disease is caused by which microorganism, what is the mode of transmission, which methods of laboratory diagnosis should be used, and what is the sensitivity to antibiotics will always be there, at least through the foreseeable future. Ever since the publication of *Bacteriology* by J. Buchanan, in 1897, numerous excellent textbooks, monographs, and exhaustive reference books covering wide-ranging topics, variously on pathogenic microbiology, clinical microbiology, medical microbiology, and microbiology of infectious diseases, have been available for quite some time now. But almost all these books are large, often more than a thousand pages long, and are consequently quite expensive for average students. During the past 40 years of his teaching carrier, one of us (SKM) has been invariably challenged by his students to recommend a concise book that is not too expensive and provides essential information. As a teacher, SKM was not aware of any such book and finally decided to do the work himself in collaboration with the youthful and well-trained infectious disease expert Dr. Dipti Agrawal.

In this era of stressful lifestyles, widespread use of immunosuppressive drugs, and enhanced exposure to environmental pollutants, it is hard to draw a sharp line between pathogenic and nonpathogenic microorganisms. If the exposure is heavy, and the host immunity is severely compromised, the list of potentially pathogenic microorganisms could be very long and beyond the scope of any concise manual of clinical microbiology. Therefore, this book aims at presenting a succinct account of the causal agents of most important and common infectious diseases with a minimal historical and introductory discussion. Illustrations, sketches, and diagrams have been kept to a minimum. This concise manual is in no way a substitute for the classics or a clinical guide book, which will always be needed for in-depth information.

Collectively, the authors bring in nearly half a century of direct experience in pathogenic microbiology and infectious diseases. Together, they have spent much of their learning careers at some of the finest institutions in the world, including Robert Koch Institute, Germany, and, in the United States, Waksman Institute of Microbiology, Rutgers, the State University of New Jersey; Boston University; University of Texas; and Baylor College of Medicine. The list of the distinguished microbiologists who mentored us includes Dr. Fritz Staib, formerly at Robert Koch Institute, the late Dr. Henry Isenberg, formerly at Long Island Jewish Hospital, and late Dr. Libero Ajello, formerly at the Centers for Disease Control and Prevention (CDC). The authors are thankful to these distinguished professionals for their guidance and also for providing some of the photographs included in the text. We are also thankful to the CDC, which in certain respects is SKM's alma mater, and also a source for many of the photomicrographs included in this manual. Finally, we express our gratitude to our family members, without whose support and encouragement this work would not have materialized. And last but not least, we express our sincere thanks to the staff of John Wiley and Sons for their patience, support, and editorial assistance.

Saroj K. Mishra, Ph.D.
Dipti Agrawal, M.D.

About the Authors

Saroj K. Mishra

Saroj K. Mishra received his Ph.D. degree in 1972 from the University of Delhi, India. He spent next 5 years as a postdoctoral fellow at Robert Koch Institute in Berlin, Germany, and at Waksman Institute of Microbiology, Rutgers, the State University of New Jersey, USA, the former being the place where the germ theory of diseases took its roots and the latter where the foundation was laid for the discovery of many powerful antibiotics. After working as a Senior Scientist at Robert Koch Institute for 5 years, Dr. Mishra joined Michigan State University as Assistant Professor and later NASA Johnson Space Center as a Senior Scientist and in charge of the Microbiology Laboratory, dedicated to support flight medicine and occupational medicine. More recently, he worked as vice president for anti-infective drug development at a pharmaceutical company and currently teaches numerous courses in microbiology at the University of Houston Clear Lake. He has authored over one hundred research papers and contributed chapters to numerous reference books. He has received numerous honors and awards, including Lady Tata Memorial, Karl-Unholz Memorial, Alexander von Humboldt Foundation, and NASA Space Act awards. In recognition of his contributions to the science of microbiology, Dr. Mishra was elected to the fellowship of the American Academy of Microbiology in 1993.

Dipti Agrawal

Dipti Agrawal received her Doctor of Medicine degree in 1996 from Boston University, followed by postgraduate training at Baylor College of Medicine, Houston, and University of Texas Southwestern Medical Center, Dallas, Texas. Since 2001, she has pursued her career as an infectious disease specialist at several hospitals and educational institutions in Houston and has written numerous research papers related to infectious diseases. Dr. Agrawal is board-certified in infectious diseases.

Chapter 1

Introduction

It is generally believed that the wars are single most destructive socio-political events with greatest impact on the society. A glance at the estimates of casualties resulting from the major wars in the 20th century can have a chilling effect:

- Number of persons killed during World War I: approximately 12 million.
- Number of persons killed during World War II: approximately 55 million.
- Combined total of persons killed in all other wars in the 20th century: approximately 1 million.

Thus, the total number of war casualties in the 20th century is estimated to be approximately 68 million.

If one takes into account all war-related deaths in the world during the past 500 years, the total would most probably be less than 100 million. In contrast, during the past century alone more than 500 million people have died of infectious diseases and nearly 5 billion have suffered from debilitating infectious diseases. Arguably, the numbers were much higher before the advent of the antibiotics era and before prophylactic vaccines became available.

Besides causing the social and emotional strain, infectious diseases profoundly affect economy and productivity of societies. There is no exact figure, but it is estimated that the worldwide health care cost during the past decade alone was several trillion dollars—much more than the total annual budget of the United States, the world's richest nation. Yet, neither nations nor societies seem to take infectious diseases as seriously as wars! Why? The answer does merit some serious consideration.

Table 1.1 is based on reports published during the past 10 years. One can very well imagine that the mortality rates were much higher in pre-antibiotics and pre-vaccine eras.

The focus of this book is on a select group of microorganisms that cause common diseases, with relatively less emphasis on the clinical aspects, though still covering the essentials. Each section deals with a group of taxonomically related microorganisms with an emphasis on the following:

A Concise Manual of Pathogenic Microbiology, First Edition. Saroj K. Mishra and Dipti Agrawal.
© 2013 Wiley-Blackwell. Published 2013 by John Wiley & Sons, Inc.

Table 1.1 Annual Mortality Due to Some of the Infectious Diseases That Mostly Affect a Large Segment of the Population of Poor Countries (Source: WHO, affiliated organizations, and authors' own experience)

Tuberculosis: Most widespread, over 2 million deaths annually
Malaria: Quite widespread, 2 to 3 million deaths annually
Cholera: Over half a million deaths annually
Typhoid: Over half a million deaths annually
Diarrhea and dysentery: Caused by a wide range of microorganisms, over 2 million deaths annually
Pneumonias and respiratory tract infections: Almost 5 million deaths annually
Mumps, measles, polio: Almost a million deaths annually
HIV/AIDS: Over 2.5 million deaths annually
Diseases classified as sexually transmitted: Nearly 0.5 million deaths

- Biology of disease-causing microorganisms (types of microorganisms)
- Natural habitats of the causal agents
- Diseases they cause and mode of dissemination
- Laboratory diagnosis
- Antibiotic sensitivity (control and prevention)

KOCH'S POSTULATE

Koch's postulate forms the very basis of the pathogenic microbiology. The causality of almost all infectious diseases is based on the postulate and theories developed by Robert Koch, who is rightly called the "father of pathogenic microbiology," and his contemporaries. Developed in the late 19th century, it has stood the test of time. The postulate can be summarized as follows:

1. A microbe suspected as the causal agent of a particular disease must be found in all subjects suffering from a similar disease but must be absent in clinical specimens from healthy individuals.

2. The suspected microorganism can be isolated from the diseased individual and grown in pure culture.

3. When this isolated suspect microbe is injected into healthy, susceptible animals (some human volunteers were also reportedly used by Robert Koch), signs and symptoms of a disease similar to the disease under investigation must develop in the infected animal.

4. The microbe cultured from the infected animal must be morphologically and physiologically identical to the strain initially isolated from the patient (in Item 1).

TERMINOLOGY

Cide: Chemical or physical agent that kills microorganisms.

- Bactericides: Agents that kill bacteria.
- Fungicides: Agents that kill fungi.
- Viricides: Chemical agents that kill viruses.
- Microbicides: Physical or chemical agents that are lethal to a broad group of microorganisms.
- Biocide: Substances lethal to all forms of life.

Communicable/contagious diseases: Diseases that are easily transmitted from person to person (e.g., tuberculosis).

Disease: Disease can be defined as a state of altered homeostasis, that is, being in a state of dis-ease. Causes and factors relating to a disease may be:

- Metabolic
- Psychiatric
- Environmental
- Infectious
- Immunological

Endotoxins: Lipopolysaccharides, produced by certain Gram-negative bacteria, for example, *Escherichia coli*.

Epidemic: Excessive frequency of a disease.

Etiologic agent: Causal agent of a disease (same as pathogen).

Exogenous: Pathogens that come from external sources (outside the body).

Exotoxins: Toxic proteins produced by a wide range of bacteria.

Facultative pathogen: A part-time pathogen, which needs to infect a host in order to complete its life cycle.

Infectious diseases: Diseases caused by microorganisms.

Mode of dissemination: Medium through which causal agents of infectious diseases are transmitted from person to person (e.g., airborne diseases).

Obligate pathogen: A pathogen that cannot live outside the host and is mostly dependant on host for the ATP.

Opportunist: Microbes that infect a host when the host's immune system is weakened.

Pathogen: A microorganism that causes a disease.

Portal of entry: The route through which a pathogen enters human body.

Reservoir: Source of infection or natural habitat of the pathogen.

Resident microbiota: Microorganisms normally present at a specific anatomical site. This is due to local physiological factors (pH, presence of iron-binding proteins, lysozyme, etc.) and the resultant species selection.

Saprobe: Microbes that grow on nonliving entities. This term mostly refers to nonpathogenic microorganisms.

Static: A chemical agent that stops cell multiplication. They can be bacteriostatic or fungistatic, for example.

Transient microbiota: Microorganisms that are not normally present at a specific anatomical site, but are introduced deliberately or inadvertently.

Virulence: Pathogenic potential of a microorganism.

Virulence factors: Morphological, physiological, or genetic traits of a microorganism that enables it to overcome the host's immune defenses. Examples include capsules, proteolytic enzymes, and toxins.

MAJOR CATEGORIES OF PATHOGENIC MICROORGANISMS

Viruses

Believed to be a bridge between the living and the nonliving, viruses have either DNA or RNA, seldom both. Their genome is surrounded by a protein coat, called capsid. Certain viruses have an envelope, often derived from the host cell membrane during lysis and release. Multiple characteristics, including type of nucleic acid, single or double strands, and presence or absence of envelope, are taken into account in the classification of viruses.

Bacteria

Bacteria are prokaryotes, which means they lack nuclear membranes and other membrane-bound organelles, such as mitochondria, Golgi apparatuses, and endoplasmic reticula. Most bacteria have a cell wall, but some are devoid of a cell wall (e.g., *Mycoplasma* spp.). Bacteria are divided into two major groups, cocci and bacilli. Cocci (singular coccus) are spherical and may occur as single coccus, as a pair called diplococcus as in the case of *Neisseria gonorrhoeae*, as a cluster as seen in the case of *Staphylococcus aureus,* or a chain of several cocci as in the *Streptococcus* spp. The bacilli (singular bacillus) are rod-shaped bacteria and they are often referred to as "rods." They exhibit a considerable variation in their size and shape. Some are straight rods, other slightly curved, and some are comma-shaped vibrios as in the case of *Vibrio cholerae.* Another variation in the shape is represented by the spiral bacteria called Spirochetes (e.g., *Treponema pallidum*, the causal agent of syphilis). Certain "evolved" forms of bacilli tend to have rudimentary filaments, as in the case of *Corynebacterium* and *Mycobacterium*, and others have a fully developed filament with true branching, as seen in the case of *Streptomyces*. Based on their reactions to Gram staining (color), both cocci and bacilli are further divided into two groups, Gram-positive (stain purple) and Gram-negative (stain red). The Gram-positive bacterial cell wall is made of a thick layer of peptidoglycan with some embedded teichoic acid. The outer layer of the Gram-negative bacterial cell wall is made of a thick layer of lipopolysaccharide, some phospholipids, and a small amount of peptidoglycan. Both groups of bacteria have aerobic (oxygen dependent) and anaerobic (oxygen independent) members. There are also several bacteria that can grow under either condition and they are called facultative anaerobes. For a detailed discussion on bacterial taxonomy, readers are referred to Bergey's Manual of Determinative Bacteriology and numerous other authoritative sources listed in the bibliography.

Fungi

Unlike bacteria, fungi are eukaryotes, which means they have nuclear membranes and membrane-bound organelle. Almost all have a cell wall, which is usually made of chitin. Based on their sexual reproduction and other structural features, fungi are divided into four major groups called divisions or phylum. Division Zygomycota can reproduce sexually and asexually and most (but not all) have no septum in their mycelium. *Mucor* and *Rhizopus* spp. are examples of zygomycetes. The second division, Ascomycota, is characterized by the production of ascospores during their sexual reproduction. The ascospores are generally housed in an enclosed structure called an ascus. An example of an ascomycete is *Pseudallescheria boydii*. The third division, Basidiomycota, produce special cells called basidia during their sexual reproduction. The sexual spores, called basidiospores, develop on the basidia. An example of a basidiomycete is *Filobasidiella neoformans*, which is the sexual stage of an important human pathogen, *Cryptococcus neoformans*. Common mushrooms are also members of division Basidiomycota. Fungi which are not known to reproduce sexually are called Fungi imperfecti (fourth division). They multiply vegetatively, but show considerable variation in their structure. As their sexual stages are discovered, they are generally categorized as either Ascomycota or Basidiomycota. A great majority of common airborne fungi, including members of the genus *Aspergillus*, *Alternaria*, and *Penicillium* are examples of Fungi imperfecti. Many fungi are also known to produce powerful toxins, such as aflatoxins, ochratoxins, aminitins, and ergot alkaloids.

Protozoa and Multicellular Parasites

Protozoa are unicellular eukaryotic microorganisms that belong to kingdom Protista. Protozoa and multicellular parasites called helminths lack a cell wall. Classification of protozoa and helminths is complex and the system is not without controversies. Therefore, the authors have chosen a simple and practical approach which is summarized in the respective chapters on unicellular parasites and multicellular parasites.

TRANSMISSION OF INFECTIOUS DISEASE (MODE OF DISSEMINATION)

Airborne (Inhalation of Bioaerosols)

A bioaerosol contains bacteria in its center, surrounded by air and a small amount of liquid, generally saliva. Bioaerosols may be produced due to sneezing, coughing, or talking. Depending on the force of sneezing or coughing, the bioaerosol-borne microorganism can travel up to several meters in air. Almost all respiratory tract infections are airborne; some can also pass from person to person through the inhalation of bioaerosols. Some of the examples of airborne infections include tuberculosis, strep throat, diphtheria, pertussis, legionellosis, influenza, and chicken pox, and a wide range of mycotic diseases such as aspergillosis, zygomycosis, cryptococcosis, histoplasmosis, and coccidioidomycosis.

Food- and Waterborne

These diseases are acquired through the ingestion of food or water contaminated with fecal bacteria or other infectious agents. The usual portal of entry is mouth or gastrointestinal tract. Examples include cholera, typhoid, botulism, shigellosis, gastroenteritis, leptospirosis, hepatitis A, and a number of parasitic diseases,

Zoonosis

The term "zoonosis" refers to a wide range of diseases that are transmitted via a vector, a carrier, or an infected animal. Such animals may also be infested with insects, which in turn harbor the pathogen. Examples are malaria, Lyme disease, rabies, plague, Rocky Mountain spotted fever, and some of the parasitic diseases.

Sexually Transmitted

These diseases are acquired through close contact, usually sexual intercourse. Examples include syphilis, gonorrhea, HIV, hepatitis B, herpes, and nongonococcal urethritis. Exchange of biological fluid through oral sex should be included in this category.

Nosocomial Infections

Nosocomial infections are infections contracted during a patient's medical care and they can include diseases contracted through fomites. Fomites are inanimate or nonliving objects, such as doorknobs, telephones, and computer keyboards, which facilitate dissemination of infectious diseases. Fomites contribute to many hospital-acquired infections. Examples include *Clostridium difficile*-associated diarrhea, urinary tract infections, and pneumonia.

UNIVERSAL PRECAUTIONS

At a minimum, universal precautions, that is, common-sense safety measures, must be followed when dealing with potentially infectious materials or persons suffering from infectious diseases. The universal precautions require an investigator to:

- Wear protective apparel (lab coat, gloves, etc.)
- Not mouth pipette
- Not talk while handling or culturing a clinical specimen
- Minimize socializing in the lab
- Not consume food or apply cosmetics in the lab
- Always disinfect work area, before and after work
- Use proper safety cabinets for the laboratory work:
  - Class I: For media preparation
  - Class II: For handling average pathogens (normally present in the area and community)
  - Class III: For handling highly pathogenic microorganisms
  - Class IV: For handling deadly pathogens (mostly used at designated labs, such as at the Centers for Disease Control and Prevention [CDC]).

Chapter 2

Host-Microbe Interactions

The human body reacts in many different ways to microorganisms. These interactions can be summarized in the following categories:

RESIDENT MICROBIOTA

All surfaces of the human body, including the skin as well as the mucous membranes that surround the inner parts of the mouth, nostrils, genitals, and gastrointestinal tract, are inhabited by a distinct set of microbial communities, which are specifically adapted to the local physical and chemical environment. Such normal microorganisms, called resident microbiota, perform extremely important roles.

1. Resident microbiota engage all available binding sites on the host cell surfaces, thus invaders have a diminished possibility of attaching to the host cell surfaces.

2. Many microbes secrete vitamins that are absorbed by the host and serve important nutritional needs.

3. Microbes also boost a form of generalized immunity against the pathogenic microorganisms that often share antigenic constituents with the resident microbiota.

4. In order to establish themselves in the host, the invading pathogen needs to establish itself and create a reservoir. Resident microbiota offer stiff competition for nutrients and may produce secondary metabolites that will discourage growth of invading pathogen.

Important Resident Microbiota

Skin

Staphylococcus epidermidis and species of *Micrococcus, Corynebacterium,* and *Propionibacterium* are the chief constituents of the resident microbiota on skin.

A Concise Manual of Pathogenic Microbiology, First Edition. Saroj K. Mishra and Dipti Agrawal.
© 2013 Wiley-Blackwell. Published 2013 by John Wiley & Sons, Inc.

Certain yeast-like fungi, such as *Candida, Malassezia*, and *Trichosporon* spp., may also be found on the skin surfaces. However, any microorganism normally present in the environment can also be isolated from the skin. These may include *Bacillus* spp. and a wide range of fungal spores. But such microorganisms are not considered constituents of the resident microbiota; they are usually referred to as "transient microbiota."

Nose and Throat

Staphylococcus epidermidis, Corynebacterium, Neisseria, and *Haemophilus* spp. are normally present in the nasal cavity and nasopharynx. *Staphylococcus aureus* can be isolated from the nose and external parts of the ears of nearly 50% of apparently healthy adults. The nose is the primary portal of entry for airborne (bioaerosol) diseases, such as tuberculosis, diphtheria, and influenza.

Mouth

The resident microbiota of the mouth include alpha hemolytic streptococci (*Streptococcus* spp.) and various species of *Corynebacterium, Neisseria, Lactobacillus,* and *Candida*. Also, a number of anaerobic bacteria are often present under the gums, just below the surfaces of the teeth. Microorganisms generally present in the air, food, and water can be also isolated from throat swabs, but they are not considered resident microbiota.

Gastrointestinal Tract

Various species of *Bifidobacterium, Lactobacillus, Bacteroides, Klebsiella, Candida, Proteus, Enterococcus,* and *Escherichia*, especially *E. coli,* are normally present in the intestine. In contrast, the stomach, due to its high acidity, offers a hostile environment to the microorganisms. A number of bacteria normally present in the intestine tract are also important opportunistic pathogens.

Vagina

Unlike the penis, which mostly harbors skin bacteria, the vagina affords a very different environment. While its high acidity and the presence of antimicrobial substances such as lactoferrin (an iron-binding protein) and S-IgA discourage the proliferation of microorganisms, the moist and often nutrient-rich conditions and partial to fully anaerobic environment favor the growth of certain microorganisms. *Lactobacillus acidophilus* is the main resident microorganism in the vaginas of women of child-bearing age. Vaginal microorganisms in prepubescent and postmenopausal stages are usually a mix of skin and intestinal microbiota. Other microorganisms that may be isolated from the vagina include species of *Streptococcus,*

Bacteroides, and *Candida*, and coliform bacteria. Elderly women may also carry *S. aureus*. A disturbed vaginal environment can predispose the host to a wide range of infections caused by common gastrointestinal microorganisms.

HOST DEFENSES

First Line of Defense

The uppermost layer of skin is comprised of dead, keratinized layer of cells. Thus, the intact skin is impervious to microorganisms, and functions as a powerful physical barrier. Most of the zoonotic diseases that are associated with insect bites result from the penetration of this barrier. Additionally, several soil and waterborne microorganisms can enter the human body through bruised or damaged skin.

Additionally, when the natural flow of the body's fluids such as urine, bile, and saliva becomes blocked due to stone formation, this may cause a buildup of bacteria that may lead to infections such as urinary tract infection, cholecystitis, and parotitis.

Second Line of Defense (Chemical Secretions)

Low pH

Low pH generally discourages bacterial proliferation. Therefore, bacterial growth in the stomach (pH 1–3) and vagina (pH 4.5–5.5) is generally low.

Chemicals

Certain iron-binding proteins found in the body's fluids discourage microbial growth by binding with free iron, which is essential for microbial growth. Some of the notable iron-binding proteins include lactoferrin (found in breast milk, semen, and the vagina) and transferrin (found in blood plasma). Lysozyme, an enzyme that dissolves bacterial cell walls, is plentiful in tears and saliva. In addition, many body secretions, such as saliva, vaginal fluids, and breast milk, contain large quantities of a secretory immunoglobulin called S-IgA or IgA2, which tends to agglutinate and immobilize invading pathogens.

Third Line of Defense (Nonspecific Immunity)

Phagocytosis

Phagocytosis by neutrophils is perhaps the most active frontier in nonspecific immunity. Neutrophils patrol the entire body and engulf and destroy nonself particles (viruses, bacteria, yeasts, etc.). Similarly activated monocytes, called macrophages, may roam the tissues or gather at the site of infection and engulf the invaders. The macrophages in the liver are called Kupffer cells and dendritic cells, those in the lungs are called alveolar macrophages, and those in the kidneys intraglomerular mesangial cells. The engulfing of nonself entities occurs through a process called endocytosis; the engulfed microorganism is contained in a vacuole called phagolyso-some. The microbial destruction in the phagolysosome may be oxygen independent or oxygen dependent. The oxygen independent mechanism involves lysozyme, lac-toferrin, and defensins. The oxygen dependent mechanism kills through the production of NADPH (reduced nicotinamide adenine dinucleotide with an extra phosphate) oxidase, myeloperoxidase, nitric oxide synthetase, hydrogen peroxide, and hydroxyl radicals.

Natural Killer (NK) Cells

NK cells are a kind of lymphocyte that is different from T- and B-lymphocytes that participate in the specific immunity. Like macrophages, NK cells too roam all over the body and seek out and destroy nonself entities that have managed to get into the bloodstream. They recognize self from nonself by sensing changes in cell surface proteins that often result from microbial infections. Malignant cells are recognized in the same manner. The process may involve a major histocompatibility complex protein (MHC I), which is found on the surface of most nucleated cells. The NK cells have specific receptors that bind to MHC I protein, if the cell is from the self (host). If the NK cell is unable to bind with the cell, it is treated as nonself. The NK cells seem to "inject" a protein called perforin, which causes lysis of the infected cells and tumor cells. The NK cells also contain Fc receptors on their surfaces. The Fc receptors can bind with the infected host cells and destroy them directly by injecting perforin.

Complements

Complements, a set of interacting serum proteins, are also components of nonspecific immunity. There are nearly 20 different types of complement components, named C1, C2, C3, C4, and so on. Their mechanisms of action may involve a classic pathway wherein the complements act on the antigen–antibody complex to destroy the pathogen by injecting perforin in the invading cells. The perforin causes holes to form in microbial cells, resulting in their death. Another pathway called alternate pathway can destroy pathogens before a specific immune response. This pathway is initiated in response to certain bacterial molecules such as lipopolysaccharides (LPS) found in the cell walls of Gram-negative bacteria. Complement C3 is cleaved into C3a and C3b, which bind with the LPS leading to the formation of an enzyme called C3bBb. Further interactions with blood protein properdin and other complements lead to the formation of a membrane attack complex (MAC), which creates holes in the plasma membrane of the pathogen, resulting in cell death.

Fourth Line of Defense (Specific Immunity)

Specific immunity is characterized by the presence of a memory system through which components of the immune system "remember" the pathogen and mount a massive attack each time the pathogen enters the body. Before any further discussion, certain related terms merit elucidation. These are cytokines and antigen. Cytokines are glycoproteins produced by numerous cells and they play an important role in specific as well as nonspecific immunity. They include interleukins (IL-1, IL-2, IL-3, etc.), interferons (IFN α, β, and IFN γ), and tumor necrosis factor (TNFα and TNFβ). The function of interleukins includes (but is not limited to) stimulation of T-cell proliferation, production and differentiation of macrophages, and growth and activation of B cells, in addition to an effect on the central nervous and endocrine systems. Properties of interferons include antiviral and antiproliferative activities and stimulation of T-cells and macrophage and NK cells. The TNFs are cytotoxic to tumor cells and have numerous other properties including mediating inflammation.

As stated earlier, the immune system recognizes foreign bodies via its ability to differentiate between self and nonself using MHT proteins. Any foreign body that is recognized as nonself is an antigen. However, in reality these are mostly glycoproteins and polysaccharide molecules commonly present on the microbial cell surfaces. Antigens have antigen determinants called epitomes, which bind to T-cell receptors or specific antibodies. A number of small organic molecules may not act as antigens by themselves but they can act as antigens if combined with a protein or similarly large molecule. Such antigens are called haptens. For example, penicillin by itself is not antigenic, but when it combines with certain serum proteins, it becomes a hapten and triggers a strong immune response in sensitive persons.

There are two lines of specific immunity, humoral and cellular immunity. It must be stated here that immunology is a highly developed branch of science and readers who want to develop an in-depth understanding of the topic should consult any of the several scholarly books in this field. **Humoral immunity** is mediated by B cells (B-lymphocytes). B cells originate and mature in the bone marrow but differentiation occurs in the lymphoid tissue. Activation of B cells can be antigen specific or T-cell dependent. The antigen presented by the macrophages stimulates T-helper cells, which in turm produce numerous cytokines that act on the B cells. The excited B cell develops into memory cells which store antigen-related genetic information, and plasma cells, which secrete specific antibodies. Antibodies, also called immunoglobulins (Ig), are essentially glycoproteins that are divided into five classes, namely IgG, IgM, IgA, IgD, and IgE. The building block of these immunoglobulins consists of two heavy chains linked together by disulfide bonds. Each heavy chain is attached to a shorter light chain via a disulfide bond. Both the heavy and the light chains have a larger constant region and a short variable region, which function as the binding sites (Fig. 2.1). A set of two light and heavy chains is also called a monomer. The IgG is a monomer and accounts for more than 80% of total serum antibodies. They neutralize toxins, opsonize pathogenic cells, and activate complements. A rise in the titer of IgG is often associated with the recovery stage of infection. IgG can also cross placenta. The IgM is a pentamer formed by connecting five monomers via

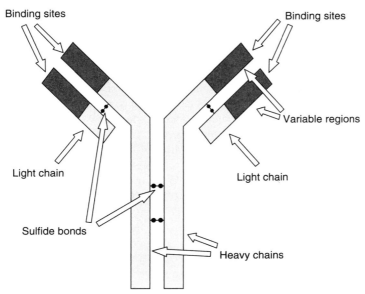

Figure 2.1.   Diagram of a monomer showing light and heavy chains, constant and variable regions, and the binding sites (IgG, IgD, IgE, and IgA1 are monomers).

disulfide bonds and a J chain. The IgM is often the first immunoglobulin to appear in the case of infection, and the titer increases as the infection progresses. IgM molecules are a very effective agglutinator and account for 5–10% of total serum antibodies. IgA can be divided into two groups, IgA1 and IgA2. IgA1, a monomer, accounts for about 5% of total serum antibodies. IgA2, a dimer (two monomers joined together with a J chain) also known as secretory IgA or S-IgA, is found on the surface of most mucocutaneous tissues lining external surfaces such as the mouth, nasal cavity, and vagina. It is also abundant in breast milk. IgD, a monomer, accounts for less than 1% of total serum antibodies. They are found on B cell surfaces and play a role in antigen recognition. IgE, also a monomer, is mostly associated with allergy. It is present in serum in very small quantities, often accounting for less than 0.05% of total serum antibodies. IgE titers rise during allergic episodes.

   Cellular immunity, the other kind of specific immunity, is mediated by T cells (T lymphocytes). T cells also originate in bone marrow, differentiate in lymphoid tissues, but mature in the thymus. The plasma membrane surface of T cells has specific receptors called T cell receptors (TCRs) that bind with the antigens. Two distinct components, an alpha polypeptide chain and a beta polypeptide chain are present on the receptor site. Part of the alpha and beta chains extend into the cytoplasm while some of their portions remain on the membrane surface. The T cells react to antigen fragments attached to the major histocompatibility complex (MHC) molecules. The eventual activation of T cells is prompted by specific molecular signaling from inside the cells. Activated T cells multiply to form more specialized subsets including T helper (T-h) and T cytotoxic (T-c) cells. In particular, T-h cells, also known as CD4+ cells, are activated by antigens presented by MHC-II molecules on the antigen presenting cells (APC). The activated T-h cells perform several functions including promoting T-cytotoxic cells, activating macrophages, and triggering production of cytokines interleukin-2, interferon gamma, and tumor necrosis factor alpha by mediating inflammation. T-h cells also stimulate antibody production and play a critical role in allergic response. The T-c cells, also called CD8+ cells, do the actual and direct destruction of the pathogen cells or host cells containing pathogens such as viruses or any other intracellular pathogen. Activation of T-c cells may occur when they interact with APC (generally the macrophages or dendritic cells). Activated T-c cells destroy the target cells either via the perforin pathway or through the CD95 pathway. In the perforin pathway, the binding of T-c cells to target cells induces cytoplasmic granules to move to the part of the cytoplasmic membrane that is in contact with the target cell. The granules then fuse with the plasma membrane and release perforin and granzymes. Perforin polymerizes the target cell membrane to produce pores. The granzymes then enter through the pores and destroy the target cells via a process called programmed cell death. During the course of CD95 pathway, the activated T-c cells increase expression of Fas ligands, which are a specific protein. Fas ligands interact with the transmembrane Fas protein receptors found on the surface of the target cells. As a result, target cell apoptosis ensues. Two other subsets of T-cells are T suppressor (T-s) and T memory (T-m) cells. The T-s cells assist in the functioning of the T-h cells and are believed to play a role in the host's allergic response.

Chapter 3

Antibiotics and Other Chemotherapeutic Agents

Technically, the antibiotic era began with the discovery of penicillin by Sir Alexander Fleming in 1929. However, its development could occur only during World War II. By that time, an energetic soil scientist, Dr. Selman Waksman, had established a school of soil microbiology in New Jersey's Rutgers University. Focusing on soil-borne aerobic actinomycetes, his group started a systematic program that lead to the discovery of streptomycin, an antibiotic credited with saving lives of millions of tuberculosis patients all over the world. Since then, his group at Rutgers as well as his students in various educational and industrial research laboratories went on to discover thousands of antibiotics, which include almost all the powerful drugs, such as tetracycline, erythromycin, chloramphenicol, amphotericin B, and vancomycin. Traditionally, the term "antibiotics" has been used for the antimicrobial agents derived from microorganisms. Since a number of antibiotics currently in use are actually synthetic, the term antibiotics has become synonymous with antimicrobial agents used for the treatment of infectious diseases.

CLASSIFICATION OF ANTIBIOTICS

From a structural perspective, antibiotics can be divided into several major categories.

Penicillin

Initially isolated from the mold *Penicillium* spp., penicillin is characterized by the presence of a β-lactam ring in the core structure (Fig. 3.1). Penicillin is, therefore, also called a β-lactam antibiotic. The β-lactam ring is sensitive to the enzyme β-lactamase, produced by a wide range of bacteria. In an attempt to develop a more effective drug, over time, the original penicillin (Penicillin G) has been repeatedly

A Concise Manual of Pathogenic Microbiology, First Edition. Saroj K. Mishra and Dipti Agrawal.
© 2013 Wiley-Blackwell. Published 2013 by John Wiley & Sons, Inc.

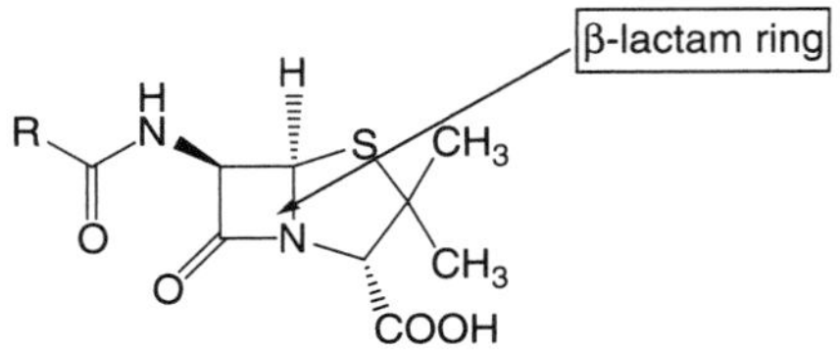

Figure 3.1.   Core structure of the penicillin molecule showing β–lactam ring.

modified. Newer penicillins include Penicillin V, Ampicillin, Methicillin, and Ticarcillin. But the β-lactamase problem still persists. Penicillin is easily absorbed and apart from the anaphylactic shock noted in some sensitive individuals, its side effects are minimal. Greater research is needed to overcome its β-lactamase-related deficiencies. Its mechanisms of action involve blocking the reaction that leads to peptidoglycan cross links and the subsequent formation of the cell wall. Without the cell wall, bacteria have poor chance of survival.

Cephalosporins

Cephalosporins are produced by the mold *Cephalosporium* spp. They are also characterized by the presence of a β-lactam-like ring, but cephalosporins are relatively more resistant to classic β-lactamase, though they are sensitive to a different kind of β-lactamase. Advances in research on cephalosporins have led to the development of four generations of this antibiotic. They have a broad spectrum, and are often used in penicillin-sensitive individuals. Their mechanism of action is similar to that of penicillin.

Polypeptides or Glycopeptides

Polypeptides are a family of powerful antibiotics that include vancomycin and bacitracin. Vancomycin is produced by *Streptomyces oreintalis*, and bacitracin by *Bacillus subtilis*. Vancomycin is mostly active against Gram-positive bacteria including methicillin-resistant *Staphylococcus aureus* (MRSA). Vancomycin is also considered a last-resort drug in managing a number of infectious diseases. Its mechanism of action involves blocking the transpeptidation reaction by binding to D-alanine terminal sequence on the pentapeptide portion of the peptidoglycan and the eventual inhibition of cell wall biosynthesis.

Aminoglycosides

These are characterized by the presence of amino sugars linked by glycoside bonds (Fig. 3.2). They include streptomycin, neomycin, and gentamicin produced by *Streptomyces griseus, S. fradiae*, and *Micromonospora purpurea*, respectively. Newer agents include tobramycin and amikacin. For several decades, streptomycin was the only antibiotic effective against *Mycobacterium tuberculosis*, the principal causal

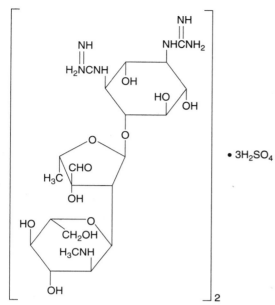

Figure 3.2.   Structure of streptomycin.

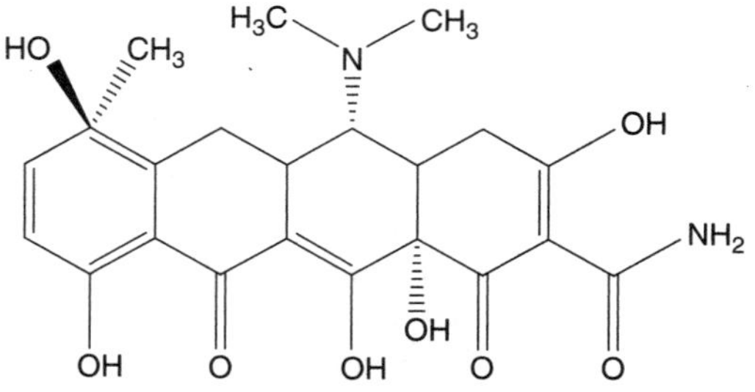

Figure 3.3.   Core structure of tetracycline.

agent of tuberculosis. Aminoglycosides are bactericidal and also effective against Gram-negative bacteria. Due to widespread resistance in *Mycobacterium* strains, the usefulness of streptomycin has been greatly, though not totally, diminished. The mechanism of action of this group of antibiotics involves blocking protein synthesis in bacterial cells by binding with the 30S subunit of the ribosomes. Aminoglycosides can be toxic, with severe side effects leading to kidney failure and hearing loss.

Tetracyclines

Original tetracycline was isolated from a strain of *Streptomyces aureofaciens*. It is characterized by the presence of four (tetra) interconnected rings (Fig. 3.3). Minor

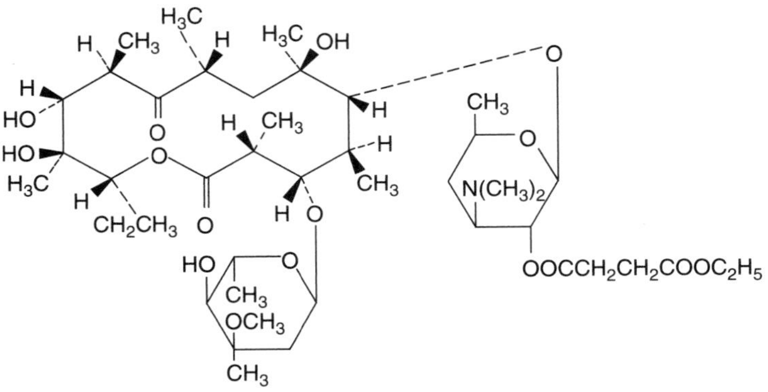

Figure 3.4.   Structure of erythromycin.

changes can be noted in different derivatives. For example, doxycycline has an extra
OH group. The tetracycline group of antibiotics is bacteriostatic, hence an active
immune system is essential in order to successfully treat an infection. These are
broad spectrum antibiotics and useful in the treatment of infections by Gram-positive
and Gram-negative bacteria, *Rickettsia*, *Chlamydia*, and *Mycoplasma*. Their mechanism of action is similar to that of aminoglycosides.

Macrolides

These are characterized by the presence a lactone ring linked to one or more sugars
(Fig. 3.4). Erythromycin, produced by *Streptomyces erythraeus*, was the first commonly prescribed macrolide antibiotic. It is a broad spectrum antibiotic effective
against Gram-positive bacteria, some Gram-negative bacteria, *Legionella*, and
Mycoplasma spp. Azithromycin and Clarithromycin are newer macrolide drugs
which have surpassed erythromycin in their usage. Like tetracyclines, the macrolides
are also bacteriostatic antibiotics. Their mechanism of action involves binding to the
23S ribosomal RNA of the 50S ribosome subunit to block protein synthesis.

Quinolones

Quinolones are synthetic antibiotics (Fig. 3.5). The best-known examples are ciprofloxacin, levofloxacin, and moxifloxacin. These are broad spectrum antibiotics and
active against enteric bacteria, as well as against *Haemophilus* and *Neisseria* spp.,
and also show a varying degree of activity against *Streptococcus pneumoniae*
and *Pseudomonas aeruginosa*. The mechanism of action includes inhibition of
DNA replication by blocking bacterial topoisomerase. Since bacterial topoisomerases are different from mammalian topoisomerases, the mechanism of action is fairly
selective.

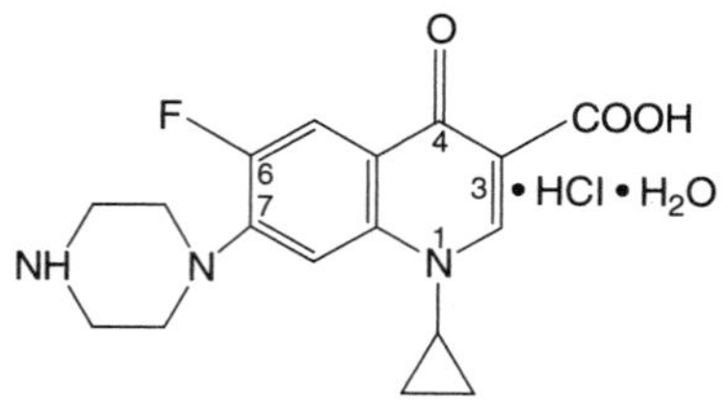

Figure 3.5. The core structure of ciprofloxacin.

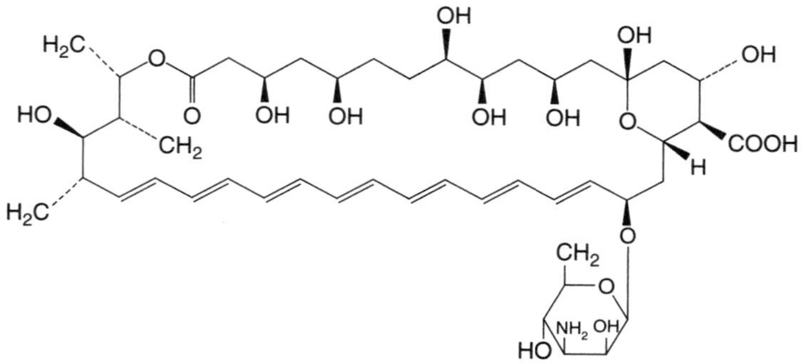

Figure 3.6. Structure of amphotericin B (note the alternate double bonds at the bottom portion of the molecule).

Polyenes

Produced by *Streptomyces* spp., polyenes are characterized by the presence of several alternate bonds in fused benzene rings (Fig. 3.6). Based on the number of alternate bonds, they are further classified as tetraene, pentaene, heptaene, and so on. The best known example is amphotericin B, which is a heptaene. Amphotericin B is often used for the treatment of invasive mycotic diseases, including invasive aspergillosis, zygomycosis, histoplasmosis, and coccidioidomycosis. Drug resistance to amphotericin B is not common. However, it is highly toxic to host and is not absorbed by the system. The mechanism of action involves forming a complex with ergosterol in fungal plasma membrane, which results in membrane leakage. Other well-known polyenes include nystatin and candicidin.

Metronidazole

Metronidazole is a nitroimidazole compound that is also known as flagyl. It is a synthetic antibiotic with a rather small molecule. It is effective against several anaerobic bacteria including *Gardnerella vaginalis, Clostridium difficile,* and a number of pathogenic protozoa, such as *Giardia lamblia, Entamoeba histolytica,*

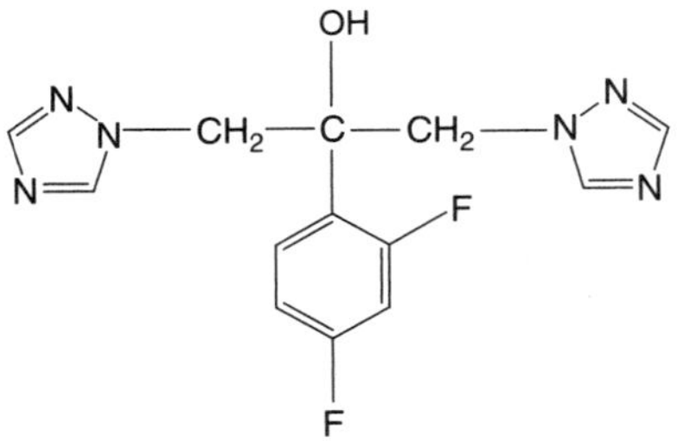

Figure 3.7.   Core structure of fluconazole.

and *Trichomonas vaginalis*. It is not active against Gram-positive bacteria. The mechanism of action involves the initial reduction of metronidazole molecules by protein cofactors from microaerophilic bacteria (e.g., *Bacteroides* spp.). The reduced metronidazole molecule cuts DNA molecules in a random manner, thus causing death of the cells.

Azole Derivatives

Azoles are relatively a new class of antifungal agents. Their discovery and development has been primarily prompted by the toxicity of amphotericin B, the only therapeutic agent available for the clinical management of systemic mycotic diseases. Currently they account for the largest number of antifungal drugs used for the management of a wide range of mycotic diseases, from superficial to systemic infections. Azole-rings are typically characterized by an N-linked methyl group formed by added halogenated phenyls or similar groups (Fig. 3.7). Azoles are basically fungistatic. The mechanism of action involves blocking the synthesis of ergosterol, a vital component of the fungal cytoplasmic membrane. Several imidazoles interfere with cytochrome peroxidase and catalase activities, causing an increase in the level of hydrogen peroxide in the cells. Based on the number of nitrogen atoms in the ring, azoles can be divided into two groups, the imidazoles and triazoles. The imidazoles have two nitrogen atoms in the ring and include clotrimazole, miconazole, econazole, and ketoconazole. Of these, miconazole is mostly used in topical preparations for the management of cutaneous and mucocutaneous candidiasis. In contrast, the triazoles have three nitrogen atoms in the ring and include fluconazole, itraconazole, and terconazole. Fluconazole and itraconazole are used orally for the treatment of some of the systemic mycotic diseases.

Mebendazole

Mebendazole (methyl 5-benzoyl-2-benzimidazolecarbamate) is a synthetic agent with broad spectrum antihelminthic properties. The core structure consists of an imidazole ring linked to a benzene ring (Fig. 3.8). The mechanism of action involves

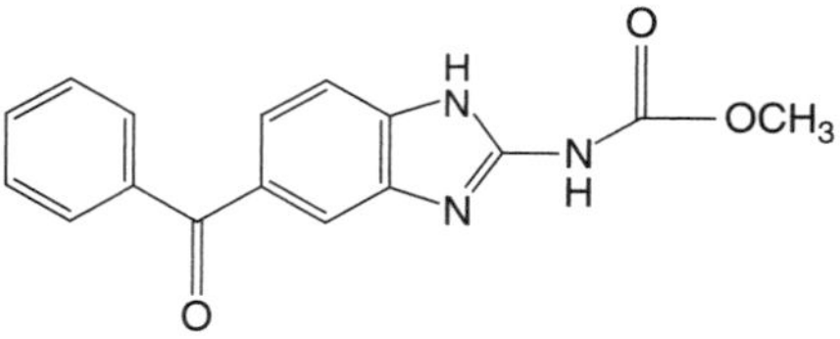

Figure 3.8.   Core structure of mebendazole.

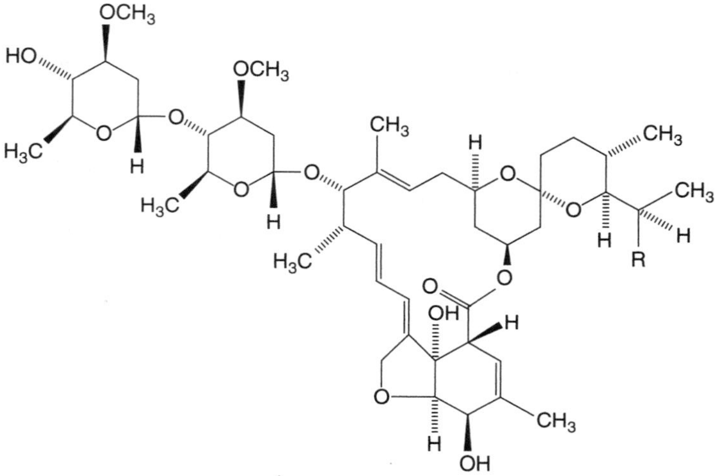

Figure 3.9.   Structure of ivermectin.

the interruption of glucose uptake, resulting in the termination of ATP production and the subsequent death of the helminths. In addition or alternatively, mebendazole binds with the tubulins and interferes with the functioning of the microtubules by blocking the assembly of tubulin dimers into tubulin polymers. As a result, motility of the helminths is seriously impaired.

Avermectins (Ivermectin)

Avermectins, commercially marketed as ivermectin, are macrolide-like compounds (Fig. 3.9) produced by the bacterium *Streptomyces avermitilis*. In a classic sense it is an antibiotic that has antihelminthic properties and is widely used in agriculture to control nematode infections. In humans, it is primarily used against filarial parasites. Its mechanism of action involves interfering with gamma-amino butyric acid (GABA) synapses in the nervous system in the helminths, resulting in their paralysis and removal from the host. Ivermectin is also known to interfere with reproduction in adult females. A single dose of ivermectin is considered enough to eliminate microfilaria for at least six months. In addition to the treatment of filariasis, ivermectin is effective against several other helminths including *Ascaris, Trichuris*, and *Enterobius* spp.

SUMMARY OF THE MECHANISMS OF ACTION

Most antimicrobial agents hit a specific target in the microbial cell. Examples of cellular targets and effective antibiotics are the following:

- **Cell wall**: Penicillins, cephalosporins, bacitracin, and vancomycin.
- **Protein synthesis**: Mostly work by blocking 30S and 50S subunits of the microbial 70S ribosome. Examples include tetracyclines, chloramphenicol, aminoglycosides, and erythromycin.
- **DNA**: There are several ways certain antibiotics can attack microbial DNA, either by directly damaging its integrity or by interfering with its replication. Examples include fluoroquinolones, rifampin, metronidazole, and sulfas.
- **Cytoplasmic membrane**: Certain antibiotics damage microbial cytoplasmic membranes. Examples include polyenes, polymyxin B, and azoles.

Chapter 4

Antiseptics and Disinfectants

The terms "antiseptic" and "disinfectant" are often confused and misused in microbiology and medicine. Typically, "antiseptic" refers to an agent used to minimize, destroy, or remove microbial population on a living surface, such as the skin of a person who needs to be prepared for injection or a surgical procedure. A disinfectant, on the other hand, is a substance used to eliminate or minimize microbial presence on an inanimate surface, such as a work bench, glassware, or surgical instruments. It is noteworthy that both antiseptics and disinfectants can either be a microbicide or a microbistatic. In this chapter, we will treat the two entities together under the banner of control of microbial population. The microbial population in or on a surface, material, or product can be controlled, minimized, or eradicated either by physical means or by chemical means, some of which are summarized below.

PHYSICAL CONTROL OF MICROORGANISMS

Heat

Traditionally, heat application has been the most preferred means of controlling microbial population. This is achieved by any of the three means, namely, incineration, dry heat, and moist heat. Incineration involves the use of very high temperatures, often (but not always) in a closed device, and is used for the disposal of medical waste and other biological material suspected of harboring dangerous microorganisms. The process kills all forms of all microorganisms. Dry heat, perhaps least efficient of the three methods, requires the application of a high temperature for an extended period of time. For example, a conventional convection oven will need at least 2 hours at 180°C to kill all microorganisms including endospores. This process is occasionally used for the sterilization of glassware and other heat-stable objects. In contrast, moist heat, by permitting faster penetration of heat into the objects, is more efficient. Boiling a material in water at 100°C for 5–10 minutes is sufficient to kill vegetative forms of most microorganisms. However, endospores do not die at this temperature. Raising the temperature to 121°C under moist conditions,

A Concise Manual of Pathogenic Microbiology, First Edition. Saroj K. Mishra and Dipti Agrawal.
© 2013 Wiley-Blackwell. Published 2013 by John Wiley & Sons, Inc.

which is achieved by the use of an autoclave at a pressure of 15 pounds per square inch (psi), can destroy all living organisms including endospores in 15 minutes. Autoclaves are often used in laboratories, hospitals, and pharmaceutical industries for the sterilization of media, glassware, surgical instruments, and other products that do not denature at this temperature and pressure.

Pasteurization is another form of heat application commonly used by beverage manufacturers to minimize microbial population in a product. This process is effective in reducing only vegetative forms of the microbial population. It does not destroy endospores and its usefulness even in destroying viable *Mycobacterium tuberculosis* cells is questionable. There are two methods of pasteurization, low temperature hold (LTH) and high temperature short time hold (HTST). The LTH requires holding the products at 62.8°C for 30 minutes and the HTST works by holding the product at 71.7°C for 15 seconds. Both processes yield identical results. Pasteurization has played a great role in reducing a number of infectious diseases earlier associated with the consumption of milk and other beverages.

Filtration

Filtration involves physically removing microorganisms from a liquid or gas. This is achieved by passing the product through a membrane with a pore size < 0.2 μm in diameter. This size can block most bacteria, fungi, and protozoa, but not viruses. However, since viruses are intracellular obligate parasites, they are of a minimal concern perhaps with the exception of hepatitis A and a few similar viruses. This method is mostly used for the sterilization of heat-sensitive products. The membrane filters generally used for this purpose are made of cellulose acetate, cellulose nitrate, polycarbonate, teflon, or any other suitable material. The advent of high-efficiency particulate air (HEPA) filters has made it economically feasible to filter a large volume of gases. HEPA filters remove particulates that are larger than 0.3 μm in size. Since most microorganisms are larger than this size, they are easily removed. However, certain bacteria, such as mycoplasma, rickettsia, and chlamydia, which are less than 0.3 μm in size, and most viruses (except smallpox and Ebola viruses) can pass through HEPA filters. HEPA filters are currently in use for air handling in laminar flow safety cabinets and in other settings that require a germ-free environment.

Radiation

All forms of life, including viruses, which are at least technically not "living organisms," are sensitive to radiation. Types of radiation that can destroy or seriously damage microorganisms and viruses are ultraviolet light and ionizing radiation. Carefully controlled radiation is used for the elimination of microorganisms and viruses from certain products and environment. The two types of radiation are described below.

Ionizing Radiation

Ionizing radiation triggers loss of electrons from the atoms. The common sources of ionizing radiation are gamma rays, having a wavelength of 10^{-3} to 10^{-1} nm, and X-rays, which have a wavelength of 10^{-1} to 10^{-2} nm. Ionizing radiation damages hydrogen bonds, double bonds, and ring structures of the molecules. Cell death and virus inactivation occurs due to destruction of nucleic acids (DNA and RNA). Most microorganisms and viruses die at an exposure of less than 1 Mrad (megarad, a unit of the measurement of absorbed radiation). However, endospores of *Clostridium botulinum*, polio, and vaccinia viruses can resist up to 1.5, 3.8, and 2.5 Mrad, respectively. Ionizing radiation is used for the sterilization of heat-sensitive materials or products that cannot be sterilized by chemical means. Laboratory petri dishes, disposable glass and plastic wares, and other products commonly used in healthcare facilities are sterilized using ionizing radiation. Ionizing radiation has a very high penetrating power and it does not leave any residue.

Ultraviolet Light

Ultraviolet light, having a wavelength of less than 400 nm, has a poor penetrating power. Therefore, ultraviolet light is mostly useful only for surface sterilization. Exposure to ultraviolet light having a wavelength of 260 nm for 1 minute is often sufficient to damage microbial DNA. Cell damage mostly results from mutation caused by formation of thiamine dimers in the DNA. Excellent results have also been obtained by exposure to ultraviolet light having a wavelength of 340 nm even though DNA does not sufficiently absorb ultraviolet light at this wavelength. Ultraviolet light is mostly used in safety cabinets and other closed environmental systems for killing microorganisms on the surfaces and in the ambient air in a small enclosed space.

CHEMICAL CONTROL OF MICROORGANISMS

Phenolic Compounds

Formerly popularly known as carbolic acid, phenol has been used as a common disinfectant by physicians and surgeons for decades. Joseph Lister, who is generally considered the father of antiseptic surgery, championed the use of carbolic acid for disinfecting surgical instruments. For decades phenol was considered to be the gold standard for determining effectiveness of a given disinfectant. However, this product is no longer in use because of its toxicity. Currently, several phenol derivatives including o-phenyl phenol, hexylresorcinol, hexachlorophene, and chloroxylenol are in use as antiseptics and disinfectants. The widely used household disinfectant Lysol is a mixture of phenolic compounds. Hexachlorophene is a popular antiseptic because of its long lasting residual effects. It is effective at a concentration of 1%–3%. Phenolic compounds act by denaturing proteins and disrupting the plasma membrane of the microbial cell.

Alcohols

Alcohols, especially ethyl alcohol (ethanol) and isopropyl alcohol (isopropanol), are bactericidal and fungicidal at a concentration of 65%–85% in water (volume by volume). However, they are not effective against endospores. A 70% concentration of ethanol or propanol is used as an antiseptic and less commonly as a disinfectant. Isopropyl alcohol is generally preferred over ethyl alcohol because it does not have the strong odor that is typical of the latter. The disadvantage of alcohols is that they evaporate rapidly, are highly inflammable, and irritate mucous membranes. Alcohols act by denaturing proteins and by dissolving lipids in the cytoplasmic membrane.

Halogens

Chlorine and iodine are most commonly used agents for microbial control. Chlorine compounds are generally used for disinfecting municipal drinking water and recreational water (swimming pools). Chlorine added to relatively clean water at a concentration of 0.5 mg/L is usually sufficient to kill most microorganisms with the exception of *Bacillus anthracis* and *Entamoeba histolytica*, which can withstand a concentration up to ten times higher. It works best at neutral pH. Chlorine reacts with water to form hypochlorous acid, which actually is the active form. With a few exceptions noted above, almost all microorganisms die within 30 minutes. Chlorine has a residual effect but its corrosive nature can harm sensitive components. Iodine and iodophors are useful antiseptics. Under certain conditions iodine is preferred over chlorine for water treatment, albeit in a limited quantity. The effective concentration of iodine in an aqueous solution is 1%. Tincture iodine, once a popular antiseptic, is a mixture of 2% iodine in 70% ethanol. It is still one of the best antiseptics for use on skin. Iodophors on the other hand are water soluble, stable, and nonstaining. Iodophors are now widely used in hospitals for preoperative skin cleaning and also as a disinfectant in hospitals as well as in laboratories. Iodine also has a residual effect, but tends to stain objects that come in contact with it. Halogens are powerful oxidizing agents and damage microbial cells by the oxidation of the sulfdryl group of enzymes.

Cationic Detergents (Quaternary Ammonium Compounds)

Quaternary ammonium compounds (quats), such as benzalkonium chloride, have lately emerged as popular antiseptics in clinical practice. Unlike to alcohol, which causes a burning sensation when applied to mucocutaneous surfaces, benzalkonium chloride is safe. It is often used for swabbing the urethral opening and vaginal areas in preparation for sample collection or surgical procedure. The effective concentration is 0.1%–0.2% in aqueous solution. It inactivates microbial cells by damaging phospholipids in their cytoplasmic membrane.

Hydrogen Peroxide

A 3% solution of hydrogen peroxide is used as antiseptic. At this concentration it is lethal to most microorganisms. Vaporized hydrogen peroxide is occasionally used for the decontamination of safety cabinets, operating rooms and other enclosed places that need to be disinfected. Under certain circumstances it is a preferred disinfectant because it does not leave behind any residue or undesirable byproduct. The end products are oxygen and water, both of which are harmless. Hydrogen peroxide kill a wide-range of microorganisms including spore formers and it works at temperatures ranging from 4 to 80°C, without causing any material damage. Like halogens, hydrogen peroxide is also a powerful oxidizing agent. It produces hydroxyl-free radicals which impair proteins and DNA molecules.

Ethylene Oxide Gas

Ethylene oxide (EtO) gas is used for disinfecting temperature-sensitive equipment at a large scale. Items such as disposable petri dishes, syringes, components of heart or lung machines, sutures, and catheters are sterilized using this gas. Its effective concentration is 0.5–0.7 g/L. EtO is quite effective in sterilizing packaged products because it rapidly penetrates packaging materials including plastic wrapping. The materials to be sterilized are placed in a chamber similar to an autoclave with instruments to control EtO concentration, temperature, and humidity. A complete sterilization is achieved in 5–8 hours at 38°C or in 3–4 hours at 54°C at a relative humidity of 40%–50%. Ethylene oxide is an explosive gas; therefore it is generally marketed as a mixture containing 15%–20% EtO in carbon dioxide. It is also highly toxic, a eye and skin irritant, highly inflammable, and carcinogenic. EtO kills microorganisms including spores by combining with cell proteins.

Aldehydes

A 3%–8% solution of formaldehyde is used for preserving tissues for histopathological examination. It is also extensively used in zoology laboratories for preserving small animals and animal organs for demonstration to students. A 37% solution (formalin) is occasionally used for vapor sterilization of a small room or laboratory suspected of contamination with pathogenic microorganisms. Another aldehyde, glutaraldehyde, is used as a high-level disinfectant at a concentration of 2%. It is often used for preserving tissues in preparation for histopathological examination. The disadvantages with aldehydes are their strong pungent odor; irritation of mucous membranes, especially the eyes and upper respiratory tract; and poor penetration. Formalin is also a carcinogen. Both glutaraldehyde and formaldehyde are alkylating agents.

Heavy Metals

A number of heavy metals, including gold, silver, copper, mercury, and zinc, have been in use as disinfectants or as antiseptics. A 1% silver nitrate solution has been widely used to prevent bacterial eye infection. Copper sulfate is generally used as an algicide in water supplies. Zinc is antifungal and is still used in ointments for topical applications. Mercury compounds, such as merthiolate (1:10,000 dilution), are used mostly in laboratories for preserving serum and other products meant for long-term storage. Another mercury salt, mercurochrome, was widely used as an antiseptic for several decades, but now its usage has been largely discontinued due to the fear of mercury toxicity. Gold, though too expensive to be used as a day-to-day disinfectant, has, nevertheless, powerful antimicrobial properties. Most heavy metals act by inactivating and precipitating vital proteins in microbial cells.

Dyes

Several dyes have strong antimicrobial properties. During the pre-antibiotics era, gentian violet solution was used for the treatment of thrush and yeast infections (vulvovaginitis) with excellent success. It is still recommended when everything else fails due to drug resistance. Other dyes with known antimicrobial properties include trypan red, malachite green, brilliant green, and certain acridine dyes. Dyes have a long-term residual effect, but stain the areas of application. Their mechanism of action involves interactions with nucleic acids. Most but not all dyes are bactericidal.

Ozone

Ozone is a strong oxidizing agent that reacts readily with most organic matters, and it is active against a wide range of microorganisms. It is more effective at lower temperatures. A minimum relative humidity of 60% is needed for disinfecting surfaces. It is generally used as a disinfectant in bottled drinking water. For that purpose, it is superior to chlorine because it is a stronger oxidizing agent and not much affected by the pH. Ozone is also preferred over chlorine for sewage treatment because of the risk of later reacting with organic molecules to form chlorinated compounds. It has little or no residual effect.

Chapter 5

Gram-Positive Cocci

BACTERIAL TAXONOMY (AN OVERVIEW)

The classic approach to the classification of bacteria is based on size and shape, later aided by reaction to Gram stain. As advances in microscope-making were made and simple biochemical tests became available, gradually the system became more refined. The science of taxonomy matured with the availability of the electron microscope and advances in molecular biology. Muller may have been the first biologist to attempt to classify bacteria in late 18th century, but his attempts were limited by lack of understanding of bacteria and crudeness of microscopes. Cohn, in late 19th century, made further advances, but serious attempts to classify bacteria were first made only in 20th century. Bergey's *Manual of Determinative Bacteriology* was first published in 1923 and it instantly became the foundation of and the most authoritative source for bacterial taxonomy. Aided by inputs from the American Society for Microbiology, international societies for bacterial taxonomy and nomenclature, and the *International Journal of Systemic Bacteriology*, Bergey's manual has been greatly refined and its scope enlarged. Its 9th edition was published in 1994. Another remarkable contribution to bacterial taxonomy was made by the publication of Bergey's *Manual of Systematic Bacteriology*, a five-volume set that examines bacterial taxonomy in greater detail. Students interested in bacterial taxonomy are encouraged to consult these highly authoritative sources.

Bacterial taxonomy, as it is, remains a complex branch of science that is not free from controversies. Several microbiologists have become famous simply by changing the names of certain species, including the names of some of the well-established species, often on trivial grounds. Authors respect modern advances but prefer a traditional approach. The tools currently available to a microbiologist for the delineation of species range from gene probes to old fashioned fermentation tests using Durham tubes. However, most modern pathogenic microbiology labs use a semi- or fully automated system, such as the Vitek or Biolog systems, for the identification of bacteria and yeast-like fungi. For the sake of convenience, the authors of this book have used a simple approach to group clinically important bacteria based on their Gram reaction, shape, gas requirements, and some simple physiological

A Concise Manual of Pathogenic Microbiology, First Edition. Saroj K. Mishra and Dipti Agrawal.
© 2013 Wiley-Blackwell. Published 2013 by John Wiley & Sons, Inc.

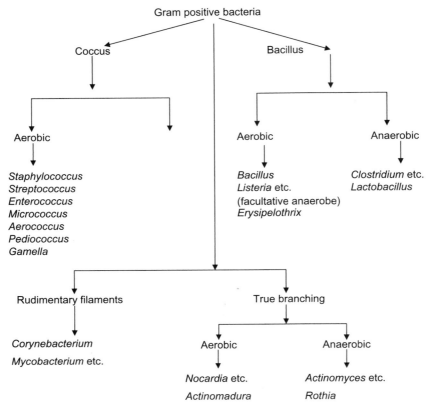

Figure 5.1.   Schema for the grouping of Gram-positive bacteria of medical importance.

tests. An example of the basis for grouping Gram-positive bacteria is illustrated in the schema below (Fig. 5.1). Similar schema for grouping Gram-negative bacteria are illustrated in the relevant chapters.

CLINICALLY IMPORTANT GRAM-POSITIVE COCCI

From the perspective of pathogenic microbiology, species belonging to the following three genera are of clinical importance: *Staphylococcus* spp., *Streptococcus* spp., and *Enterococcus* spp. Members of a fourth taxa, *Micrococcus* spp., are closely related, but they are mostly harmless constituents of normal skin microbiota. Major differences between *Staphylococcus* and *Micrococcus* spp. are summarized in Table 5.1.

Staphylococcus Species

Staphylococci are endogenous (found within the host) bacteria, commonly associated with the skin, nose, ear, and throat. *Staphylococcus aureus* is the most important

Table 5.1   Some Differentiating Characteristics of *Staphylococcus* and *Micrococcus* spp.

Tests	*Staphylococcus*	*Micrococcus*
Acid from glycerol	Positive	Negative
Resistance to:		
lysozyme	Resistant	Sensitive
bacitracin	Resistant	Sensitive
Oxidase	Negative	Positive

pathogen. Other species that may be incriminated in diseases include *S. epidermidis*, *S. saprophyticus*, *S. haemolyticus*, and *S. lugdunensis*, which are otherwise considered constituents of normal skin microbiota. Species of *Staphylococcus* that may colonize the human body but are not known to cause diseases are *S. saccharolyticus*, *S. warneri*, *S. hominis*, *S. auricularis*, *S. xylosus*, *S. simulans*, *S. cohnii*, and *S. pasteuri*. Approximately 30% of the general population and more than 50% of medical professionals are carriers of *S. aureus*, mostly in their nostrils or skin.

Diseases

Clinical conditions triggered by *S. aureus* and related species are summarized below.

- Soft tissue infections such as folliculitis and cellulitis, abscess formation, and toxic shock syndrome
- Bacteremia and endocarditis
- Osteomyelitis
- Pneumonia and empyema
- Toxin-related food poisoning
- Infections involving eyes, nose, throat, urethra, and vagina (in elderly women)

Virulence Factors

Toxins including hemolysins, leukocidins, enterotoxins A through E, and toxic shock syndrome toxin (TSST) may be produced by strains of *S. aureus* in human hosts. Extracellular enzymes including coagulase, catalase, hyaluronidase, lipases, and proteinases are also known to cause damage to the host.

Laboratory Diagnosis

Blood agar is an excellent isolation medium. Mannitol salt agar is a selective medium. Staphylococci are aerobic; they grow well at 35°C with visible colonies within 18–24 hours. The colonies are smooth, spherical, and opaque with a golden hue. Many strains cause β-hemolysis on blood agar.

Table 5.2   Important Differentiating Characteristics of *Staphylococcus* spp. Other Than *S. aureus*

Test	*S. epidermidis*	*S. hominis*	*S. haemolyticus*	*S. warneri*	*S. cohinii*
Nitrate reduction	Positive	Variable	Variable	Variable	Negative
Thermonuclease	Negative	Negative	Negative	Negative	Negative
Alkaline Phosphatase	Positive	Negative	Negative	Negative	Negative
Urease	Positive	Positive	Negative	Positive	Negative
Resistance to novobiocin	Negative	Negative	Negative	Negative	Positive
Carbon source and acid production:					
D-Trehlose	Negative	Variable	Positive	Positive	Positive
D-Mannitol	Negative	Variable	Variable	Variable	Variable
D-Cellubiose	Negative	Negative	Negative	Negative	Negative
Maltose	Positive	Positive	Positive	Variable	Variable

Taxonomy

The distinctive characteristics of *S. aureus* are the following:

- Positive coagulase test
- Positive thermonuclease test
- Fermentation of mannitol

Gene probes are also available for the identification of staphylococci. Differentiation with other *Staphylococcus* spp. requires elaborate physiological tests. Some of the relevant characteristics are summarized in Table 5.2.

Antibiotic Sensitivity

Staphylococci used to be very susceptible to penicillin. However, over time and with increased usage of beta lactam antibiotics, resistant strains have emerged as a major challenge. A group of such resistant strains are called MRSA (methicillin resistant *Staphylococcus aureus*). A sensitivity test can be helpful. Currently, vancomycin is a mainstay of therapy for serious infections but resistance to it is increasing worldwide. Newer agents such as daptomycin and linezolid are also being used as well as new cephalosporins that are the only beta lactams to cover MRSA, which include ceftaroline and ceftobiprole.

Streptococcus Species

Streptococci are catalase-negative, Gram-positive cocci that form short or long chains *in situ*. Several members of the genus *Streptococcus* constitute normal microbiota of mouth and upper respiratory tract. These can also be isolated from gastrointestinal tract and female genitourinary tract. *Streptococcus mutans, S. salivarius, S. sanguis,* and *S. mitis* are some of the nonpathogenic species, that are collectively referred to as Viridans streptococci.

Diseases

There are three pathogenic species: *Streptococcus pneumoniae, S. pyogenes,* and *S. agalactiae.*

Streptococcus pneumoniae, also referred to as pneumococcus, is known to cause pneumonia, otitis media (midear infection), sinusitis, meningitis, and bacteremia (septicemia). Most strains of *S. pneumoniae* are nonhemolytic. The common serotypes isolated from clinical specimens are referred to as serotype 6, 14, 18, and 19. This bacterium can also be isolated from the respiratory tract of apparently healthy individuals. Infections occur when pneumococci get into the lungs or bloodstream. Under normal circumstances nonspecific immune components, especially phagocytes and macrophages, keep these bacteria under control. Pneumococcal pneumonia is often accompanied by high fever and chill and the symptoms may be mistaken for viral infection. The cough may be productive and blood tinged. Children and the elderly are more prone to pneumococcal pneumonia. Ear infections are more common in children, but the sinusitis can affect adults as well. Bacteremia is more serious form of infection and not an uncommon complication in pneumonia. Bacteremia can also lead to endocarditis and meningitis.

Streptococcus pyogenes, also called Group A streptococcus, causes infections involving the upper respiratory tract (pharyngitis or strep throat) and mucocutaneous tissues, as well as skin infection, scarlet fever, and muscle infection (hence the term "flesh-eating bacteria"). The incubation period of strep throat infection is very short and the symptoms may include sore throat, fever, and headache. The cervical lymph nodes may be affected. In some cases, strep throat infection may lead to scarlet fever, characterized by body rashes that may heal in about 1 week. The skin infection may begin with exposure of bruised skin to *S. pyogenes* either through direct contact or through inanimate objects (fomite). Pus-filled vesicles develop and eventually rupture and crust over. Another form of skin infection, called necrotizing fasciitis, involves deeper subcutaneous tissue and results in a severe destruction of the muscles. The disease often develops into a systemic infection and eventual death in many cases. Infection with this bacterium is also known to lead to secondary, or post streptococcal glomerulonephritis (PSGN), which is characterized by inflammation due to immune complex formation in glomeruli, resulting in hematuria (blood in urine) and proteinuria (high protein concentration in urine). Another secondary consequence of *S. pyogenes* infection is rheumatic fever. This usually occurs 2–3 weeks after acute pharyngitis with *S. pyogenes*. It may involve the heart (resulting in damaged heart valves), joints (multiple joint arthritis), central nervous system, and skin.

Streptococcus agalactiae, also called Group B *Streptococcus*, is normally present in the gastrointestinal tract. Isolation of this bacterium from the respiratory tract of apparently healthy individuals is not uncommon. In addition, mostly due to poor personal hygiene, *S. agalactiae* can contaminate the vaginal area without causing any clinical symptom in the woman. This bacterium can cause neonatal pneumonia, bacteremia, and meningitis. Infections in adults are usually less severe, limited to fever and malaise. Cases of upper respiratory tract infection due to Group B *Streptococcus* in those who perform oral sex on women are not uncommon.

Virulence Factors

Streptococcal virulence factors can be divided into two categories:

1. **Extracellular toxins**: These include Streptolysin-S, Streptolysin-O, streptokinase, and pyrogenic exotoxins. Streptolysin-S is mostly produced by members of Group A, C, and G. Streptolysin-O destroys hemoglobin and suppresses chemotaxis. It can also cause lysis of leukocytes and erythrocytes. Streptokinase causes lysis of blood clots and facilitates the spread of bacteria.

    Streptococcal pyrogenic exotoxins A, B, and C are believed to be responsible for the rashes in scarlet fever and for the symptoms of toxic shock syndrome. A rise in the distribution of streptococcal pyrogenic toxin-producing strains is believed to be associated with the rise in Group A streptococcal invasive infections.

2. **Cell-associated factors**: The cell-associated factors include M protein, which helps bacterium in evading host's immune system, and lipoteichoic acid (mostly associated with *S. pneumoniae*), which facilitates adherence. Capsules as seen in *S. pneumoniae* protect the pathogen from the host's defenses and also from the effects of antibiotics and other chemicals.

Lab Diagnosis

Appropriate clinical specimens, such as throat swab, sputum, or blood, must be obtained under aseptic conditions. Often, Gram staining can provide useful clues. Pathogenic streptococci require complex media for growth. Best results are obtained with blood agar, which also helps in the detection of hemolysis. Brain heart infusion agar is also good. Incubation is done at 35°C in aerobic conditions for 24 hours. Colonies of *S. pyogenes* are very small, translucent, and surrounded by a clear, large, and nearly transparent zone caused by β-hemolysis. Colonies of *S. agalactiae* are relatively larger, flatter, and creamy in appearance. The β-hemolysis zone in the case of *S. agalactiae* is much smaller. Colonies of *S. pneumoniae* are initially dome-shaped and mucoid or watery with none to slight α-hemolysis. Colonies of Viridans streptococci are much smaller and opaque with a strong α-hemolysis.

Immunological tests including enzyme linked immunosorbent assay (ELISA) and agglutination tests aimed at group-specific carbohydrates are useful in detecting Group A streptococci directly in the throat swab. A number of commercial latex agglutination kits are available for the rapid detection of *S. pyogenes, S. pneumoniae,* and *S. agalactiae.*

Taxonomy

Several physiological characteristics are taken into account for the delineation of the *Streptococcus* spp. In addition, the patterns of hemolysis and Lancefield classification of Beta hemolytic streptococci are helpful.

1. Alpha (α) hemolysis: Characterized by partial destruction of erythrocytes accompanied by greenish/brownish tinge around the colony (e.g., Viridans streptococci).
2. Beta (β)-hemolysis: Characterized by total destruction of erythrocytes and a clear zone around the colony (e.g., *S. pyogenes*).
3. Gamma (γ) hemolysis: Characterized by an absence of hemolysis (e.g., *S. pneumoniae*).

Like *Enterococcus* and *Lactobacillus* spp., streptococci are catalase negative (do not split H_2O_2 into O_2 and H_2O).

The Lancefield classification of beta hemolytic *Streptococcus* spp. is shown in Table 5.3.

Viridans Streptococci

Viridans streptococci include hemolytic and nonhemolytic species that constitute normal microbiota of mouth and upper respiratory tract. Hemolytic species are known to produce only α-hemolysis. A short list of important Viridans streptococci based on CDC classification is given below:

- *Streptococcus mutans*
- *S. salivarius*

Table 5.3 Lancefield Classification of Beta Hemolytic *Streptococcus* spp. Based on Antigenic Properties

Species	Lancefield group	Main habitat
Streptococcus pyogenes	Group A	Humans
Streptococcus agalactiae	Group B	Humans, cattle
Streptococcus equi	Group C	Horses
Enterococcus faecalis	Group D	Humans, animals

- *S. sanguis*
- *S. mitis*
- *S. anginosus*

Viridans streptococci can be identified on the basis of the utilization of mannitol, sorbitol, arginine, and hydrolysis of esculin. *Streptococcus mutans* strains are almost always positive with mannitol, sorbitol, and esculin, while *S. mitis* are always negative. *Streptococcus sanguis* is positive with arginine and esculin while *S. salivarius* is positive only on esculin. Most Viridans streptococci are negative in urease test.

Antibiotic Sensitivity

Initially, streptococci were quite sensitive to penicillin. Cephalosporins and erythromycin are used in penicillin-sensitive hosts. In some cases, a combination of ampicillin and an aminoglycoside is indicated. Cases of rheumatic fever may need long periods of medication. Other effective agents include vancomycin and levofloxacin. However, instances of resistance to any of these antibiotics, including newer macrolides (azithromycin and clarithromycin), have been noted in recent years. Pneumococcal vaccine is effective in the prevention of *S. pneumoniae* infections, although patients may be susceptible to strains not included in the vaccine.

GRAM-POSITIVE COCCI RELATED TO *STREPTOCOCCUS* SPECIES

Aerococcus viridans

Involvement of *Aerococcus viridans* in human cases has been reported, albeit sparingly. This bacterium is normally present in the environment and is often encountered as a constituent of resident microbiota in humans. It has been occasionally reported as the causal agent in cases of bacteremia, meningitis, endocarditis, septic arthritis, and urinary tract infection.

Gamella Species

Human cases of infection by *Gamella* spp. are rare. Isolation of one of the *Gamella* spp., *G. hemolysans*, has been reported from the upper respiratory tract and incriminated in subacute endocarditis. Like *G. hemolysans*, another species, *G. morbillorum*, too has been occasionally isolated from the genitourinary and respiratory tracts, blood, and abscess swabs. It has been, at least on rare occasions, incriminated in the cases of endocarditis and brain abscess. It must be noted that because of their similarities, these are perhaps frequently misidentified as Viridans streptococci. Therefore, the possibility of *Gamella* spp. being more frequently involved in the causation of human diseases cannot be ruled out.

Pediococcus Species

Eight species of the genus *Pediococcus* have been isolated from clinical specimens, though most of them sparingly. Of these, *P. acidilactici* is of a relatively greater clinical significance. It has been reported as the causal agent of septicemia. It has been also isolated, with uncertain causality, from the cases of leukemia and neutropenia. In addition, *Pediococcus acidilactici* has been noted in several cases of bone marrow transplants undergoing digestive system decontamination therapy with gentamicin, vancomycin, and colistin.

Enterococcus Species

Members of the genus *Enterococcus* were previously called Group D streptococci. However, based on detailed physiological and genetic tests, these are now placed in a different genus, *Enterococcus*. Enterococci are normally present in the human intestine. *Enterococcus faecalis* is the main pathogenic species that accounts for a large number of cases of urinary tract infection (UTI). Cases of bacteremia have also been attributed to this bacterium. Endocarditis can be seen as a secondary complication in cases of bacteremia and can be difficult to cure.

Laboratory Diagnosis

Enterococcus faecalis grows well on blood agar and chocolate agar. A selective medium especially formulated for this pathogen is also available. Most strains of *E. faecalis* are catalase negative. Enterococci can be differentiated from other Group

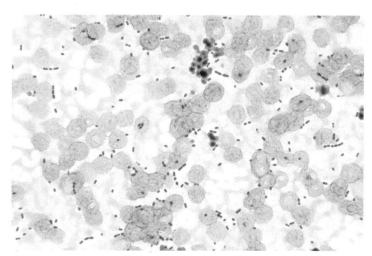

Figure 5.2.   A Gram-stained smear from a clinical specimen showing Gram-positive cocci of *Enterococcus*.

D streptococci by their ability to hydrolyze pyrrolidonyl arylamidase and grow in media containing 6.5% sodium chloride. Enterococci are Gram-positive cocci. They are often seen singly, but occasionally form a short chain (Fig. 5.2).

Antibiotic Sensitivity

Enterococcus faecalis is notoriously resistant to most antibiotics. A combined therapy using an aminoglycoside and ampicillin or vancomycin has proven effective in some cases. Some of the newer agents being used for MRSA, such as linezolid and daptomycin, also have efficacy against *enterococcus.*

Chapter 6

Gram-Positive Bacilli

Important pathogens in this group of bacteria are restricted to three genera that include aerobes as well as anaerobes:

- *Clostridium* spp. (anaerobic)
- *Bacillus* spp. (aerobic)
- *Listeria* spp. (aerobic to facultative anaerobic)

Most Gram-positive rods grow well on blood agar. However, selective media are available in some cases, for example, cycloserine–cefoxitin–fructose–egg yolk agar (CCFA) for *C. difficile*.

CLOSTRIDIUM SPECIES

Clostridia are anaerobic, spore-forming (sporogenous), Gram-positive rods. They are widely distributed in nature and many are common soil-borne bacteria. Their ability to produce spores enables them to survive under adverse environmental conditions and also affords them greater protection from chemical and physical agents. Recent genetic studies seem to challenge the conventional classification of the members of this genus. Depending on the school of thought, number of species assigned to genus *Clostridium* varies from 20 to more than 100. Of these, four species, namely *C. tetani, C. botulinum, C. perfringens and C. difficile*, are most important pathogens. Some of the physiological differences among the clinically significant *Clostridium* spp. are summarized in Table 6.1.

Clostridium tetani

Clostridium tetani is anaerobic, Gram-positive, endospore-forming bacillus that is normally present in soil and dust. Its isolation from fecal matter is not uncommon.

A Concise Manual of Pathogenic Microbiology, First Edition. Saroj K. Mishra and Dipti Agrawal.
© 2013 Wiley-Blackwell. Published 2013 by John Wiley & Sons, Inc.

Table 6.1   Some Differentiating Characteristics of *Clostridium* spp.

Species	Mannose fermentation	Indole Test	Lacithinase test	Lipase
C. botulinum	Negative	Negative	Negative	Positive
C. difficile	Positive	Negative	Negative	Negative
C. perfringens	Positive	Negative	Positive	Negative
C. tetani	Negative	Variable	Negative	Negative

Disease

Clostridium tetani causes tetanus, a dreaded, painful, and often fatal disease. The incubation period may range from 1 to 54 days (generally 6–15 days). Widespread use of antitetanus vaccines in United States and other industrialized countries has greatly reduced number of cases, but tetanus is still fairly common in rest of the world. There are several clinical variations of tetanus but the generalized form of the disease is most common. The organs involved include masseter and back muscles and the autonomic nervous system. The most important symptoms are persistent spasm of back muscles, cardiac arrhythmia, irregular blood pressure, sweating, dehydration, and a condition called Risus sardonicus, or twisted facial muscles.

Virulence Factor

Clostridium tetani produces two types of toxins: tetanospasmin, a spasmogenic neurotoxin, and tetanolysin, a hemolytic toxin. In addition, it produces endospores, which resist most antimicrobial substances produced by components of the immune system.

Laboratory Diagnosis

Tetanus is usually diagnosed on the basis of clinical symptoms and patient's history of exposure to soil, dust, or rusted metal. Cultures are seldom attempted and not always successful. Gram staining can provide useful information in early stages (Fig. 6.1).

Antibiotic Sensitivity

Tetanus toxin is irreversibly bound to tissue so supportive care is of primary importance. However, tetanus immune globulin combined with metronidazole should be administered. Active toxin production is limited by wound management. Penicillin is also an acceptable alternative to metronidazole. The best protection against tetanus is obtained by vaccination with three doses of tetanus toxoid, followed by booster doses every 10 years.

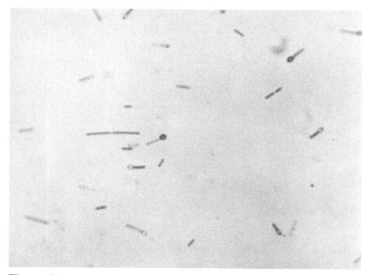

Figure 6.1.   A Gram-stained smear of *Clostridium tetani* showing Gram-positive vegetative rods, some with a terminal endospore.

Clostridium botulinum

Like *C. tetani, C. botulinum* is an ubiquitous bacterium that is commonly isolated from soil and stagnant water. Eight serotypes (A, B, C_α, C_β, D, E, F, and G) have been reported.

Disease

Clostridium botulinum causes botulism, which is a paralytic disease. Three clinical forms of botulism are known: food borne infection, caused by ingestion of toxin-infested food, infant infection, caused by ingestion of food contaminated with bacterium, and wound infection, resulting from bacterial contamination. Symptoms of food-borne infection include blurred vision, constipation, abdominal discomfort, and muscle weakness. Death occurs due to respiratory failure.

Virulence Factor

Most strains produce botulinol (botulinum toxin), an extremely potent toxin that acts on the neuromuscular junction. The bacterium has an additional advantage due to its ability to produce endospores.

Laboratory Diagnosis

In the cases of botulinum food poisoning, fecal material is cultured on enriched media. Since such materials are heavily contaminated with a wide range of

Figure 6.2.   A Gram-stained smear prepared from a culture of *Clostridium botulinum*, showing Gram-positive rods with endospores.

fast-growing bacteria, it is better to first prepare a saline suspension of the fecal matter and then leave it for 10 minutes in a water bath held at 80°C. This process kills vegetative forms of most contaminating bacteria. The sample is then inoculated on an enriched medium such as egg yolk agar. The incubation is done anaerobically at 35°C. Gram-positive rods with endospores are easily seen in Gram-stained smears (Fig. 6.2).

Antibiotic Sensitivity

A trivalent botulinum antitoxin has been used with success. Penicillin should also be administered to those with wound botulism. Metronidazole is also believed to be an effective agent and is an acceptable alternative in those allergic to penicillin.

Clostridium perfringens

Like other Clostridia, *C. perfringens* is ubiquitous and commonly present in soil and water. It is also frequently isolated from human feces. The endospores enable this bacterium to survive under unfavorable environmental conditions.

Disease

Clostridium perfringens causes gas gangrene, involving invasion of healthy muscles surrounding the infected site. The condition can be life-threatening. Infections occur as a result of contamination of fresh wounds. Bullae and necrosis may develop in the infected wound, which gradually becomes dark colored.

Virulence Factors

Strains of *C. perfringens* are known to produce several toxins. These are usually characterized as α, β, and λ toxins. In addition, phospholipases and proteinases produced by some strains probably play a role in tissue destruction and gas production.

Antibiotic Sensitivity

Proper wound care is required and the wounds can heal normally in due course. Medication with penicillin may be required in serious cases.

Clostridium difficile

A small number of *C. difficile* bacteria are normally present in the human intestinal tract. Disease develops when the normal bacterial population is destroyed by the prolonged use of broad-spectrum antibiotics.

Disease

Clostridium difficile causes diarrhea, mostly following long-term therapy with broad-spectrum antibiotics. The clinical condition is called pseudo membranous colitis (PMC) and may lead to dehydration, electrolyte imbalance, toxic megacolon, and colon perforation. These bacteria are frequently isolated from hospital floors, bedpans, and equipment, and found in the intestine of 10%–20% of healthy persons. A hypervirulent epidemic strain, referred to as NAP-1, has become more prevalent over the recent years. It is believed to be responsible for much of the increasing virulence and mortality being seen with this infection. It effects the expression of toxins A and B, as noted in the virulence factors section.

Virulence Factors

Most strains of *C. difficile* produces two toxins (A and B). Toxin A is enterotoxin and B is cytopathic. Toxin B is believed to be more significant in causing the clinical disease.

Laboratory Diagnosis

Tissue culture, polymerase chain reaction (PCR), latex agglutination and ELISA are helpful diagnostic tools. Bacterial culture can be made on cycloserine–cefoxitin–fructose–egg yolk agar. Toxin formation needs to be demonstrated in order to distinguish active infection from colonization. In the clinical setting, the most useful tests are the ELISA for toxins A and B and PCR for the gene for toxin B.

Antibiotic Sensitivity

Effective agents include metronidazole and oral vancomycin, which is not absorbed. Replenishing some of the bacteria (e.g., *Lactobacillus*) normally found in intestinal tract can be helpful. A new agent, fidaxomicin, was approved by the FDA in 2011 for the treatment of *C. difficile*–related medical conditions.

LACTOBACILLUS SPECIES

Lactobacilli are not pathogenic bacteria, but a section has been added here because they are also Gram-positive, anaerobic, asporogenous bacilli. They play a positive role in human health, especially in the health and well being of women by protecting vaginal surfaces from colonization by harmful bacteria. Genus *Lactobacillus* has several species. Many are found in the gastrointestinal tract and vagina. *Lactobacillus acidophilus* ferments glycogen into lactic acid, which causes an acidic environment in the vagina, thus excluding other bacteria that prefer neutral pH. Surprisingly, *L. acidophilus* is mostly absent in the vagina of prepubescent and postmenopausal women, suggesting a possible connection with estrogen.

BACILLUS SPECIES

Member of the genus *Bacillus* are aerobic, sporogenous, rod-shaped bacteria. The term "bacillus" refers to a specific bacterial genus, and also refers to a specific bacterial shape, that is, the rod shape. The genus *Bacillus* has more than 40 species, but only one is an important pathogen, *B. anthracis*. Another member of this genus, *B. cereus*, is also of clinical significance. Physiological differences between *B. anthracis* and *B. cereus* are summarized in Table 6.2.

Table 6.2   Growth-Related Differences between *B. anthracis B. cereus and B. thuringiensis*

Characteristic	*B. anthracis*	*B. cereus* and *B. thuringiensis*
Thiamine requirement	Positive	Negative
Hemolysis on sheep blood agar	Negative	Positive
Poly D-glutamic acid capsule	Positive	Negative
Lysis by gamma phage	Positive	Negative
Motility	Negative	Positive
Growth on chloral hydrate agar	Negative	Positive
String-of-pearls test	Positive	Negative

Bacillus anthrasis

Bacillus anthracis, also known as anthrax bacillus, causes the deadly disease anthrax in humans. The bacterium can be found in soil. Cattles in developing countries often suffer from skin anthrax. The infections are acquired either through inhalation or through direct contact with the skin of infected animals or their hides. Anthrax is almost nonexistent in the United States and other industrialized countries. Increased awareness in recent years is primarily due to its potential use as a biological weapon by terrorists. The bacterium produces highly resistant endospores and it can be easily grown in culture.

Disease

Three clinical types of anthrax—pulmonary, cutaneous, and gastrointestinal—are generally recognized. The infection may spread via blood to meninges. The pulmonary form, relatively less common than the other two, is an airborne infection. The symptoms include fever, cough, headache, and vomoting. In advanced cases, lymph nodes are enlarged and symptoms of meningeal infection may develop. This form of infection can take a rapid course, leading to death in a very short time, often within 3 days. The cutaneous form involves necrotizing ulcers on skin which may or may not take a systemic course. The lesions become increasingly dark with time. Mortality is uncommon in the cases of skin anthrax. Gastrointestinal anthrax is aquired through ingestion of contaminated food. The bacterium may invade the upper intestinal tract, cecum, and ileum. Its symptoms include nausea and malaise. Hematogenous dissemination leads to systemic infection which is usually fatal.

Virulence Factors

Bacillus anthracis posses a capsule made of poly D-glutamic acid that helps it evade host defenses. Most *B. anthracis* strains produce tripartite toxins which comprise three proteins, named LF, PA, and EF.

Laboratory Diagnosis

Specimens (blood, pus, and other body fluids) must be collected aseptically. The bacterium grows well on blood agar at 35°C. Phenylethyl alcohol agar is used for isolation from heavily contaminated specimens. *Bacillus anthracis* is a fast-growing bacterium. Class III or Class IV safety cabinets should be used when handling suspected specimens or cultures. Fresh isolates are strongly Gram-positive and endospores can be readily seen. The Gram reaction weakens as the cultures age.

Taxonomy

Due to the presence of a capsule, *B. anthracis* is easily differentiated from other members of the genus *Bacillus*. Important delineating characteristics of main species of the genus *Bacillus* are summarized in Table 6.3.

Table 6.3   Physiological Differences between Some Common Species of the Genus
Bacillus

Species hydrolysis	Penicillin	Lecithinase	β-hemolysis	Starch	Capsule
B. anthracis	Sensitive	Variable	None	None	Present
B. cereus	Resistant	Positive	Present	Positive	Absent
B. subtilis	Sensitive	Negative	Variable	Positive	Absent
B. sphaericus	Sensitive	Negative	Negative	Negative	Absent

Antibiotic Sensitivity

Effective agents include doxycycline, ciprofloxacin, and chloramphenicol.

Bacillus cereus

Bacillus cereus is a common soil-borne bacterium and is easily isolated from the
air. Contamination of cereals, vegetables, meat, and cooked food commonly occurs.

Disease

Food poisoning is the most common form of *B. cereus* infection. The disease is
acquired via the ingestion of contaminated food. Symptoms include nausea, vomiting, diarrhea, and abdominal pain. Cases of ocular infection are occasionally reported.
Such infections often follow a traumatic injury and simultaneous contamination with
the soil.

Virulence Factors

Two enterotoxins, one heat stable and the other heat labile, have been reported.
Cereolysin and a phospholipase are believed to cause tissue damage.

Laboratory Diagnosis

Bacillus cereus is easily grown on most media, especially on blood agar. Incubation
is done at 35°C under aerobic conditions. *Bacillus cereus* strains show peritrichous
flagella, which can be visualized by the use of Leifson flagella stain (Fig. 6.3).

Antibiotic Sensitivity

Antibiotic therapy is not indicated in most cases. Serious conditions can be treated
with clindamycin, vancomycin, gentamycin, or ciprofloxacin.

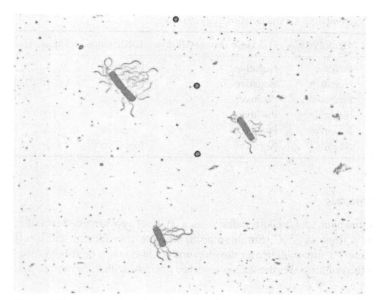

Figure 6.3. A smear from the culture of a *B. cereus* strain showing peritrichous flagella (Leifson flagella stain). See color insert.

LISTERIA SPECIES

Listeria monocytogenes

Listeria are Gram-positive, facultative anaerobes, nonsporogenous bacilli. The only species of medical importance is *L. monocytogenes*.

Disease

Listeria monocytogenes causes meningitis and septicemia and accounts for 12%–15% of all cases of meningitis in infants less than one month old or elderly persons aged more than 70 years. The bacterium is commonly present in the environment and it can grow at 4°C. Because of its ability to grow at low temperatures, the practice of storing leftover baby food including the feeding bottle in the refrigerator may play important role in infants contracting this disease. Infections occur due to consumption of contaminated food or animal product. Of the human population, 1%–5% harbors this bacterium in intestine and vagina. The possibility that some newborns acquire this infection while passing through the birth canal cannot be ruled out.

Virulence Factors

Most pathogenic strains are β-hemolytic and also produce phospholipases.

Table 6.4   Some Differentiating Characteristics of *Listeria* spp.

Species	Hemolysis	D-Xylose fermentation	L-Rhamnose fermentation
L. monocytogenes	Positive	Negative	Positive
L. innocua	Negative	Negative	Variable
L. ivanovii	Positive	Positive	Negative
L. welshimeri	Negative	Positive	Variable
L. seeligeri	Negative	Positive	Negative
L. grayi	Negative	Negative	Negative

Laboratory Diagnosis

Blood agar, Columbia agar, and colistin-nalidixic acid (CNA) agar are good isolation media. Incubation is done at 35°C. Small smooth, whitish, translucent, and moist colonies approximately 1 mm in diameter develop within 24 hours. The β-hemolysis can be seen under the colonies. Fermentation tests are positive with L-rhamnose and negative with xylose.

   Genus Listeria has six species that include *L. monocytogenes, L. innocua, L. ivanovii, L. welshimeri, L. seeligeri*, and *L. grayi*. Except for *L. monocytogenes*, all other *Listeria* spp. have been isolated from nonhuman as well as human sources. It is, therefore, important to differentiate *L. monocytogenes* from other five closely related species (Table 6.4).

Antibiotic Sensitivity

The therapeutic agents generally used for clinical management include ampicillin, gentamicin, and trimethoprim-sulfa.

Erysipelothrix rhusiopathiae

Erysipelothrix rhusiopathiae is a Gram-positive, asporogenous, catalase negative bacillus. In humans, it mostly causes localized inflammation of skin characterized by painful, red zones. The infection may spread to the blood and the vascular system, resulting in septicemia and endocarditis. The bacterium is widely distributed in nature and can be isolated from cattle, horses, pigs, mice, and turkeys, which may either be carriers or infected with *E. rhusiopathiae*. Most infections are due to occupational exposure in those such as butchers, fisherman, slaughterhouse workers, and farmers who come into close contact with these animals.

Virulence Factors

No clearly defined virulence factor has been identified but its ability to attach to heart valves and produce enzymes such as neuraminidase and hyaluronidase could possible play a role in the pathogenesis.

Laboratory Diagnosis

Erysipelothrix rhusiopathiae grows well on blood agar. Better results are obtained if the biopsied tissue is homogenized and then inoculated on the media. The incubation is done at 35°C. This bacterium is aerobic, but the growth is generally better if the incubation is done in a carbon dioxide incubator. The colonies grow slowly and may take up to 7 days to fully develop. They may present variable size and texture from rough and large to small and clear. This bacterium differs from other common Gram-positive pathogenic bacilli by its catalase negative properties.

Antibiotic Sensitivity

Erysipelothrix rhusiopathiae strains are mostly sensitive to ampicillin, clindamycin, and cephalosporins.

Chapter 7

Gram-Positive Bacteria with Rudimentary Filaments

Taxonomically, bacteria with rudimentary filaments rank between classic bacilli and the Actinomycetes, bacteria with branching filaments. They represent an important evolutionary link in microbiology. Among the bacteria with rudimentary filaments, *Corynebacterium* and *Mycobacterium* are two important pathogenic genera.

CORYNEBACTERIUM DIPHTHERIAE

Members of the genus *Corynebacterium* are aerobic to facultative anaerobic, catalase positive, and do not produce spores. They are common constituents of resident microbiota on human skin and in the mouth and upper respiratory tract. More than 16 species are recognized. Of these, *C. diphtheriae* is the most important pathogenic species. Like several other causal agents of diseases, asymptomatic carriers of *C. diphtheriae* are not uncommon.

Disease

Corynebacterium diphtheriae causes diphtheria, an acute, highly infectious upper respiratory tract infection that may eventually involve the heart and nervous system. Diphtheria occurs all over the world, especially in poor countries. But major outbreaks have also occurred in more developed countries. For example, an outbreak in 1994 in what was then the Soviet Union affected more than 50,000 persons and caused nearly 2,000 deaths. Otherwise, the widespread use of vaccines in United States and Western Europe has almost eradicated this disease. However, cases of diphtheria are still recorded each year in the United States. The incubation period can vary from 2 to 7 days. The symptoms include sore throat, malaise, inflammation of pharynx and breathing problems. In advanced cases, the heart and central nervous system (CNS) are often involved. The disease is fatal if untreated. Another species,

A Concise Manual of Pathogenic Microbiology, First Edition. Saroj K. Mishra and Dipti Agrawal.
© 2013 Wiley-Blackwell. Published 2013 by John Wiley & Sons, Inc.

Table 7.1   Some Differentiating Characteristics of Important *Corynebacterium* spp.

Species	Nitrate reduction	Urease production	Sucrose utilization	Maltose utilization	β-hemolysis
C. diphtheriae	Positive	Negative	Negative	Positive	Variable
C. ulcerans	Negative	Positive	Negative	Positive	Positive
C. striatum	Positive	Negative	Positive	Negative	Negative
C. renale	Positive	Positive	Negative	Negative	Negative
C. xerosis	Positive	Negative	Positive	Positive	Negative

C. ulcerans, can also cause diphtheria-like disease. Some of the physiological differences among clinically significant species of the genus *Corynebacterium* are summarized in Table 7.1.

Virulence Factors

An exotoxin that blocks protein synthesis in the host has been recognized as an important virulence factor.

Laboratory Diagnosis

The pathogen can be isolated on Loffler's agar (a selective medium) and blood agar. Incubation is normally done at 35°C with or without CO_2 for 24–48 hours. Since *Corynebacterium* spp. are common constituents of resident microbiota, it is important to differentiate the isolates. Gram staining of smears prepared from laboratory grown cultures mostly shows long rods, and occasionally some rudimentary filaments (Fig. 7.1).

Antibiotic Sensitivity

Corynebacterium is notorious for variable antibiotic sensitivity; therefore, it is important to perform antibiotic sensitivity test. Depending on the sensitivity, penicillin, erythromycin, tetracycline, or even vancomycin may be required. Vaccines have proven quite effective in preventing diphtheria.

MYCOBACTERIUM SPECIES

Compared with *Corynebacterium, Mycobacterium* strains tend to show a greater tendency to form rudimentary branching filaments. Such filaments easily break into bacillary bodies due to their fragility. Therefore, in smears, mostly bacilli are seen. Mycobacteria are theoretically Gram-positive, but are difficult to stain and hard to

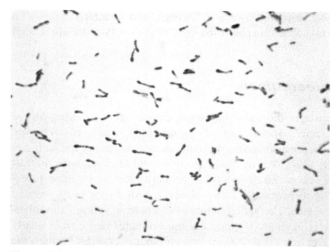

Figure 7.1.   A Gram-stained smear showing Gram-positive rods and rudimentary filaments of *C. diphtheriae.*

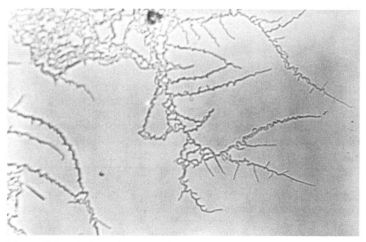

Figure 7.2.   A photomicrograph of *Mycobacterium fortuitum* (grown on tap water agar) showing filaments with rudimentary branching (source: CDC).

kill because the cells are coated with a thick layer of lipid. The best way to see the rudimentary filaments is to grow them in a slide culture or observe undisturbed growth on a nearly transparent medium using a long working distance (LWD) objective lens (Fig. 7.2). At least 50 species are known; about seven are pathogenic. Ziehl-Nelsen stain, also called acid-fast staining, is very useful in separating Mycobacteria from actinomycetes. The staining involves treating the smear with a boiling

Carbolfuchsin solution, decolorizing with acid alcohol, and counterstaining with malachite stain. Because mycobacteria can resist decolorization, they are also called acid-fast bacilli (AFB).

Mycobacterium tuberculosis

Mycobacterium tuberculosis is the most important causal agent of tuberculosis. Other species, such as *M. bovis, M. kansasii, M. gordonae, M. avium, M. fortuitum, M. phlei*, and *M. smegmatis*, are also considered pathogenic, but mostly for immune-compromised subjects. Of these, *M. fortuitum, M. phlei*, and *M. smegmatis* are fast growers. *Mycobacterium avium* and *M. bovis* are primarily known to cause tuber-culosis in birds and cattle, respectively. Tuberculosis is perhaps the most dreaded disease of a global significance. The actual number of people suffering from tuber-culosis on any given day is anybody's guess, but the estimates range from nearly two billion cases in 2000, to more than 200 million on other occasions. Perhaps the actual number will never be known but it is a fact that a large number of people suffer from tuberculosis, from a benign subclinical to active fulminating disease. It is also a fact that the disease is curable and there is no reason why it could not be eradicated from the face of the earth. Important physiological differences among clinically significant species of the genus *Mycobacterium* are noted in Table 7.2.

Disease

Mycobacterium tuberculosis is the principal causal agent of tuberculosis, a slowly progressing disease that may involve any organ including the lungs, brain, kidney,

Table 7.2 Comparison of Some Properties of Clinically Important Species of the Genus *Mycobacterium* (Modified after Howard et al., *Clinical and Pathogenic Microbiology*, Mosby, St. Louis)

Species	Colony appearance	Niacin reduction	Nitrate reduction	Urease production	Pyrozina-midase production	Percentage human cases
M. tuberculosis	Rough	Positive	Positive	Variable	Positive	90
M. bovis	Rough	Negative	Negative	Variable	Negative	2
*M. kansasii**	Variable	Negative	Positive	Variable	Negative	<1
M. gordonae	Smooth	Negative	Negative	Variable	Variable	<1
M. avium	Variable	Negative	Negative	Negative	Positive	1
*M. fortuitum***	Variable	Negative	Positive	Positive	Positive	<1
*M. smegmatis***	Variable	Negative	Positive	Variable	Variable	<1
*M. phlei***	Rough	Negative	Positive	Variable	Negative	<1

* Believed to be a common waterborne bacterium. Almost all human cases were noted in AIDS patients.
** Fast-growing bacteria that have been isolated from water and sewage as well.

and the gastrointestinal tract. Bronchopulmonary tuberculosis is the most common form of the disease. Its symptoms include cough, blood-tinged sputum, malaise, weight loss, and mild fever, which may rise in the afternoon. Radiologically, the appearance may range from pulmonary abscess to cavitary or miliary forms of tuberculosis. The incubation period can be as short as a few weeks to several years or decades. Infection can stay dormant for many years but flare up when immunity is diminished. Tuberculosis is a slow killer and it is almost always fatal if not treated. It is a classic example of airborne (bioaerosol borne) disease.

Virulence Factors

No clear-cut virulence factors are recognized except that the pathogen can grow in unactivated alveolar macrophages. Host factors, most importantly cellular immunity, play a critical role in the pathogenesis of tuberculosis. Therefore, those with HIV or on immunosuppressive drugs, as well as those with diabetes and cancer are at increased risk. Also, individuals with genetic or drug-induced defects in the interferon gamma pathway may be particularly susceptible to mycobacterial disease.

Laboratory Diagnosis

In the case of the adults, a first morning sputum every 24 hours for 3 days is examined. For children, a gastric washing is preferred. Occasionally, especially in the case of adults, a sample of bronchial aspirate is more helpful. Sputum is mixed with an equal amount of NALC (N-acetyl L-cysteine): NaOH, mixed vigorously, and centrifuged at 3,000 RPM for about 15 minutes. The sediment is streaked on Lowenstein-Jensen agar or Middle brook agar, which are selective media. Cultures are incubated for up to 3 weeks or longer at 35°C in aerobic conditions. Biopsied tissues do not require NALC:NaOH treatment. The colonies of *M. tuberculosis* appear rough and non- to slightly pigmented. Other strains may produce smooth colonies with yellowish-orange pigment. Skin test using old tuberculin (OT) or a purified protein derivative (PPD) can be helpful in making a presumptive diagnosis, if supported by clinical and radiological findings. In much of Asia, Africa, and Latin America, a positive skin test merits several considerations. Since a rather large number of subjects are exposed to the pathogen, many are tuberculin positive but devoid of any positive radiological and clinical findings. This may suggest latent tuberculosis infection. However, often these individuals have been previously vaccinated, making interpretation difficult. Newer interferon gamma release assays may help to distinguish latent tuberculosis from merely vaccinated individuals. If a person is suspected to have latent tuberculosis infection, he or she should be evaluated by an expert to determine if treatment is appropriate. Stained smears mostly show acid-fast bacilli. As stated earlier, unlike the slide cultures, in prepared smears mostly bacilli are seen as depicted in an electron micrograph in Figure 7.3.

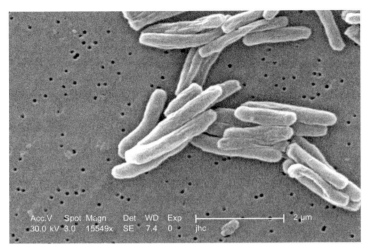

Figure 7.3.   A scanning electron micrograph of *M. tuberculosis*, 15,000× (source: CDC).

Antibiotic Sensitivity

Rifampin or ethambutol, isoniazid (INH), and pyrazinamide are quite effective in the clinical management of tuberculosis. A large number of strains are now considered streptomycin resistant. In most cases, the clinicians prefer to start with all four drugs, rifampin, ethambutol, INH, and pyrazinamide. Once the sensitivity results are available, the number of drugs may be reduced. Streptomycin, though rarely used these days, can be helpful in drug resistant cases. In order to fully eradicate the infection, a long term therapy, often ranging from 6 to 9 months to more than 1 year is required.

Mycobacterium leprae

Disease

Mycobacterium leprae causes leprosy, a highly infectious/contagious disease that mostly involves skin and the periphery. Daily showering and soap application has almost eliminated this disease in the industrialized world. However, approximately 10 million people still suffer from leprosy globally. Cases of leprosy have been reported, even in Texas, Louisiana, and California, but mostly in prison inmates. Infection is generally acquired through direct skin-to-skin contact or contact with contaminated fomites.

Laboratory Diagnosis

Most strains of *M. leprae* cannot be cultured on laboratory media. No vaccination is available. Diagnosis is often made on the basis of clinical symptoms and demonstration of acid-fast bacilli in clinical specimens (Fig. 7.4).

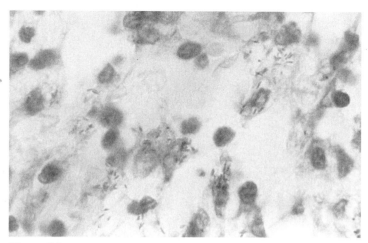

Figure 7.4.   Acid-fast bacilli in a skin biopsy from a leprosy patient (source: CDC). See color insert.

Antibiotic Sensitivity

The clinical management of leprosy is achieved by therapy with rifampin in combination with dapsone given over a long period of time. Proper personal hygiene and avoiding direct contact with infected patients can be a helpful preventive measure.

Chapter 8

Gram-Negative Cocci

For a general discussion on Gram-negative bacteria, readers are referred to the last part of this chapter. With reference to Gram-negative cocci, members of the following three genera are considered important pathogens:

- Neisseria spp.
  - ◦ *N. gonorrhoeae*
  - ◦ *N. meningitidis*
- *Moraxella* spp.
  - ◦ *M. catarrhalis*
- *Haemophilus* (It is a coccobacillus and sometimes listed among the Gram-negative bacilli)
  - ◦ *H. influenzae*

NEISSERIA SPECIES

Neisseria spp. are aerobic, Gram-negative, nonmotile, oxidase positive cocci that do not produce endospores. Members of several species of *Neisseria,* including *N. sicca, N. mucosa, N. lactamica,* and *N. flavescens,* commonly colonize mucous membranes of mouth and nasopharynx. Two species are important human pathogens. These include *N. gonorrhoeae* and *N. meningitidis.*

Neisseria gonorrhoeae

Neisseria gonorrhoeae, often referred to as gonococcus, causes gonorrhea, a disease that was quite rampant during the pre-antibiotics era. This bacterium infects only humans; no animal cases and no other natural reservoirs are known.

A Concise Manual of Pathogenic Microbiology, First Edition. Saroj K. Mishra and Dipti Agrawal.
© 2013 Wiley-Blackwell. Published 2013 by John Wiley & Sons, Inc.

Disease

Gonorrhea is a sexually transmitted disease. Isolation of *N. gonorrhoeae* from inanimate objects is occasionally reported but their role in the causation of gonorrhea is equivocal. The infection initially involves the urethra in men and cervix in women. Incubation period may range from less than 2 days to more than 1 week, and the symptoms include purulent discharges and dysuria. Disseminated gonorrhea may involve the joints, heart, meninges, eyes, and the pharynx. In women, it may spread to the fallopian tube and cause pelvic inflammatory disease (PID), which may result in ectopic pregnancy and infertility. More than 400,000 cases of PID occur annually in the United States. Gonorrhea is more common among women than men perhaps due to the fact that many female cases remain asymptomatic hence do not receive timely medical attention. Since infected females may not present any symptom, they usually play a major role in the dissemination of gonorrhea in general population. Gonorrhea in males is mostly symptomatic and easily detected.

Virulence Factors

Only a few virulence factors are known. Certain proteins, namely pilin, protein I and II, and an IgA specific proteinase, appear to play an important role in the pathogenesis of this bacterium. In order to infect, the pathogen must attach to the mucosal cells of the epithelial walls by means of fimbriae. The bacterium invades the spaces separating columnar epithelial cells, which are found in the oral pharynx, eyes, rectum, urethra, cervical opening, and external genitals of prepubescent females.

Laboratory Diagnosis

Blood agar is an excellent isolation medium, but some bacteria and yeasts are known to have inhibitory effect on *N. gonorrhoeae*. Therefore, New York City agar is preferred for the selective isolation of *N. gonorrhoeae*. The inoculated plates should be incubated at 35°C, aerobically (anaerobic incubation is also acceptable), for 24–72 hours. Microscopic examination of pus demonstrating presence of Gram-negative diplococci can be helpful (Fig. 8.1). This bacterium produces acid from glucose, but not from maltose, lactose sucrose, or fructose.

Antimicrobial Sensitivity

Initially, penicillin was the wonder drug for the treatment of gonorrhea and is credited with the eradication of this disease from most of the industrialized countries. However, during the past few decades, a large number of multiple drug resistant cases have emerged. Currently, the preferred agents include azithromycin, doxycycline, and ceftriaxone. The CDC recommends a single dose of ceftriaxone. Prevention is best achieved by the use of condoms.

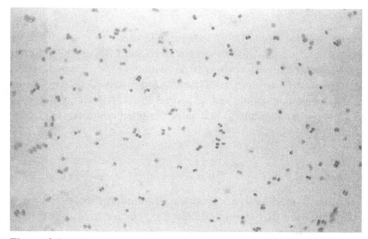

Figure 8.1.   Gram-negative diplococci seen in a smear prepared from *Neisseria gonorrhoeae* isolated from a case of gonorrhea.

Neisseria meningitidis (Meningococcus)

Neisseria meningitidis is strictly a human pathogen and is frequently isolated from throat and nasal swabs of apparently healthy individuals. Its isolation from vaginal and labial swabs has also been occasionally noted.

Disease

Neisseria meningitidis can be isolated from the upper respiratory tract of about 10% of the healthy population. The primary site of infection is the nasopharynx from where it may spread to the blood stream and cause septicemia and meningitis. Clinical symptoms may also include headache and fever. Children younger than 3 years of age are most susceptible, but cases involving adults are not uncommon. The infection is mostly acquired through inhalation of bioaerosol. Living in close quarters with infected persons is a major risk factor. The disease in untreated patients with meningococcal meningitis is almost always fatal.

Virulence Factors

Virulence factors include pili that facilitate colonization, lipooligosaccharide which contains the endotoxin moiety that is responsible for fever and other symptoms, opacity proteins, and capsular polysaccharides. Host factors such as splenectomy and terminal complement deficiency play a role in susceptibility to severe infection in a minority of patients.

Laboratory Diagnosis

Gram staining of cerebrospinal fluid (CSF) can provide clues to a presumptive diagnosis. Suitable clinical specimens, such as throat swab and CSF, can be cultured on blood agar and grown aerobically (growth is better with 3%–5% carbon dioxide) at 35°C for 1–2 days. *Neisseria meningitidis* is fastidious and sensitive to low temperature; therefore, the specimens should not be stored in a refrigerator. Colonies are very small, round, smooth, and whitish. Specially designed commercial kits for blood culture are useful.

Antibiotic Sensitivity

Penicillin and cephalosporins are effective antibiotics. Chloramphenicol is indicated in some cases but is rarely used. Rifampin and ciprofloxacin are used for prophylaxis, however, increasing resistance to ciprofloxacin is being reported in some areas. Vaccination with a tetravalent vaccine from the polysaccharide capsular antigen is very effective in preventing meningococcal meningitis in military recruits. Its effect on children is, however, questionable.

MORAXELLA CATARRHALIS

Disease

Moraxella catarrhalis, occasionally referred to as *Branhamella catarrhalis*, is an important causal agent of otitis media in children. It can also cause meningitis, endocarditis, bronchopulmonary infections, and neonatal conjunctivitis. This bacterium can be isolated from the throat swabs of apparently healthy individuals.

Laboratory Diagnosis

Blood agar is a useful medium for the isolation of *M. catarrhalis*. Incubation should be done aerobically at 35°C for 24 hours. The colonies are convex, nonpigmented, and smooth. This is a fast-growing bacterium. It is oxidase positive and β-galactosidase negative. Occasionally, *Moraxella* spp. are mistaken for *Neisseria* spp. Important differences between the two taxa are summarized in Table 8.1.

Antibiotic Sensitivity

Moraxella catarrhalis infections are often hard to treat. Most strains produce β-lactamase. The clavulanate-supplemented amoxicillin can be useful. Other effective antibiotics include tetracycline, cephalosporins, and fluoroquinolones.

Table 8.1 Some Delineating Properties of *Neisseria* and *Moraxella* spp.

Species	NY city agar	Chocolate agar	Acid from maltose	Acid from lactose
N. gonorrhoeae	Growth	No growth	Negative	Negative
N. meningitidis	Growth	No growth	Positive	Negative
N. mucosa	No growth	Growth	Positive	Negative
N. sicca	No growth	Growth	Positive	Negative
N. lactamica	Growth	Variable	Positive	Positive
N. flavescens	No growth	Growth	Negative	Negative
M. catarrhalis	Variable	Variable	Negative	Negative

Table 8.2 Some Physiological Differences between Some Clinically Significant Species of the Genus *Haemophilus*

Species	Hemolysis	Glucose fermentation	Mannose fermentation	Catalase test
H. influenzae	None	Positive	Negative	Positive
H. haemolyticus	Present	Positive	Negative	Positive
H. parahaemolyticus	Present	Positive	Negative	Positive
H. parainfluenzae	None	Positive	Positive	Variable

HAEMOPHILUS INFLUENZAE

Members of the genus *Haemophilus* are commonly found in the mouth and upper respiratory tract of humans. Ten species including (but not limited to) *H. influenzae, H. parainfluenzae, H. haemolyticus*, and *H. parahaemolyticus* are commonly isolated from human sources. These are Gram-negative coccobacilli that are often grouped together with other Gram-negative bacilli. Only *H. influenza* is of main clinical significance. It must be stated here that the term "*Haemophilus influenzae*" is a classic example of misnomer in microbiology. It is neither hemophilic nor a causal agent of influenza. It can be isolated from the respiratory tract of approximately 5% of apparently healthy persons. Some physiological differences among commonly encountered *Haemophilus* spp. are noted in Table 8.2.

Disease

Haemophilus influenzae causes acute or chronic severely invasive infection of the respiratory tract and may involve meninges, especially strains belonging to the serotype b, generally referred to as "Hib." It is a common cause of infection in children younger than 3 years of age. Prior to the introduction of vaccine, Hib accounted for nearly 20,000 cases of invasive infections in the United States annually. Infection is contracted via inhalation of *H. influenzae*-laden bioaerosol. In adults, infections involving sinuses and ears are also seen.

Virulence Factors

Haemophilus influenzae serotype b (Hib) has polyribitol phosphate capsule that helps the bacterium resist phagocytosis. In addition, pili help it with the initial attachment and colonization.

Laboratory Diagnosis

Most strains of H. influenzae do not grow well on blood agar without a feeder bacterium, which is usually *S. aureus*. *Staphylococcus aureus* strains produce a growth factor called "V factor" that helps *H. influenza* grow (Fig. 8.2). Chocolate agar, which does not need a feeder bacterium is, therefore, a preferred medium for the isolation of *H. influenza*. On chocolate agar, the colonies appear as small dewdrops. The growth conditions include incubation at 35°C and moist atmosphere with 5%–10% CO_2.

Taxonomy

As stated above, genus *Haemophilus* has many species, which are normally associated with the mouth and upper respiratory tract. A delineating scheme for some of the closely related species is depicted in Table 8.2.

Figure 8.2.   A demonstration of the use of feeder bacterium *S. aureus* for the isolation of *H. influenzae*. See color insert.

Antibiotic Sensitivity

About 20%–30% strains produce β-lactamase, hence are resistant to penicillin (ampicillin). Effective agents include cephalosporins, erythromycin, and fluoroquinolones. Vaccination is the most effective prophylaxis for children younger than 18 months of age.

AN OVERVIEW OF GRAM-NEGATIVE BACTERIA

Gram-negative bacteria are the most abundant in nature. While some are found in humans as commensals, only a few are pathogenic, and a great majority of them are

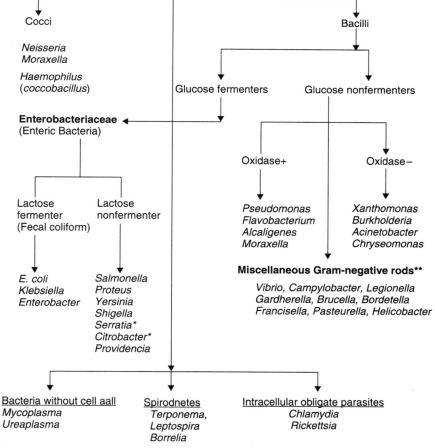

Figure 8.3. Schema for the grouping of some clinically significant Gram-negative bacteria. *Some species are lactose fermenters. **Some taxa present curved rods and some are coccobacilli (see text).

harmless to humans. Because of their abundance in nature and their metabolic and structural diversity, there has been a considerable amount of controversy in their taxonomic classification. Traditionally, Gram-negative bacteria have been placed under the division Gracillicutes in the Kingdom Monera. Prompted by recent advances in molecular genetics, the newer generation of microbiologists believes that Gram-negative bacteria are genetically too diverse to be placed under one division and resect the classic concept of high-level division of bacteria into four classes that were initially based on cell wall composition and Gram staining. There may be some strength in this argument, but molecular taxonomy is still far from being fully developed and practical. We believe that until and unless a workable taxonomic scheme based on molecular genetics is at least partially if not fully developed, the traditional system should remain in place.

The single most important component in the cell wall of Gram-negative bacteria is the abundance of lipopolysaccharide, also called LPS or just endotoxin, which plays a critical role in their pathogenicity. LPS triggers the innate immune response and production of cytokines that play a critical role in inflammation and cellular immunity. As for the taxonomy, readers are referred to *Bergey's Manual of Determinative Bacteriology* for a scholarly discussion and an in-depth classification. Figure 8.3 presents a simple plan for grouping clinically significant Gram-negative bacteria.

Chapter 9

Gram-Negative Bacilli

Asporogenous Gram-negative bacilli of clinical importance can be divided into two major groups. Glucose-fermenting, oxidase-negative, and catalase-positive members constitute one group, called Enterobacteriaceae. Several members of this group are normally present in human intestines and others are causal agents of serious infections. The second group, somewhat more heterogeneous, usually called nonfermentative Gram-negative bacilli, are glucose nonfermenters. They are widely distributed in nature and prefer aquatic habitats. However, several members of this group are frequently isolated from human sources and known to cause serious infections. A simple and practical scheme for the grouping of important pathogenic Gram-negative bacteria is depicted at the end of the previous chapter.

Members of the family Enterobacteriaceae are further divided into two groups; lactose fermenters, also called fecal coliforms or just coliforms, include *Escherichia coli*, *Klebsiella*, and *Enterobacter*. Lactose nonfermenters include many pathogenic species, such as *Salmonella typhi, Shigella dysenteriae*, and *Yersinia pestis*.

SPECIMEN COLLECTION

Clinical specimens, such as fecal matter, urine, blood, and other body fluids, should be collected aseptically and processed immediately. Otherwise, transport media should be used.

MEDIA AND LABORATORY DIAGNOSIS

Most Gram-negative rods grow on a wide range of media; some are selective, and others are nonselective (also called all-purpose media). Specific media for different groups will be described under the relevant section. Some of the media commonly used for their isolation are listed in Table 9.1.

As is evident from the information summarized in Table 9.1, MacConkey, Eosin methylene blue, and Hekteon agars have low selectivity, but Bismuth sulfite and

A Concise Manual of Pathogenic Microbiology, First Edition. Saroj K. Mishra and Dipti Agrawal.
© 2013 Wiley-Blackwell. Published 2013 by John Wiley & Sons, Inc.

Table 9.1 Growth and Colony Characteristics of Some Members of the Family Enterobacteriaceae on Some Selective Media (Modified after Howard et al., *Clinical and Pathogenic Microbiology*, Mosby, St. Louis)

Species	MacConkey agar	Eosin-methylene blue agar	Hektoen agar	Bismuth sulfite agar	Brilliant green agar
Escherichia coli	Flat, red or pink	Metallic sheen	Yellow-orange	No growth	No growth
Klebsiella pneumoniae	Dark pink, mucoid	Colorless	Yellow or green	No growth	No growth
Enterobacter spp.	Pink, slightly mucoid	Purple	Yellow-orange	No growth	No growth
Serratia spp.	Colorless to slightly pink	Lavender or colorless	Color-less	No growth	No growth
Proteus mirabilis	Colorless	Colorless	Color-less	No growth	No growth
Salmonella spp.	Colorless	Colorless	Green or blue	Green-black	Pink-white
Shigella spp.	Colorless	Colorless	Green or blue	No growth	No growth
Yersinia pestis	Colorless or peach	Colorless or purple	Salmon	No growth	No growth

brilliant green agars are highly selective. In addition, *Salmonella* spp. and *Escherichia coli* can be differentiated from others by their ability to produce "black butt" on triple sugar iron agar (not included in Table 9.1).

Incubation is done for 18–24 hours at 35°C under aerobic or anaerobic conditions. *Serratia marcescens* can grow at 1–5°C and *Escherichia coli* does well at 45°C. In general, members of Enterobacteriaceae are easy to isolate and identify. Diagnosis of certain infections may require serologic test.

Occasionally, it is better to use nonselective media, especially for isolation from blood and sterile body fluid (e.g., cerebrospinal fluid). Sometimes use of enrichment broth, such as Gram-negative (GN) broth and Salenite-F (SF) broth, is desirable, especially for *Salmonella* and *Shigella* spp.

ENTEROBACTERIACEAE

As stated above, members of the family Enterobacteriaceae are Gram-negative, nonspore-forming bacilli that are oxidase negative, catalase positive, and ferment glucose with acid production. They reduce nitrate to nitrite. Important members of Enterobacteriaceae, commonly known as enteric bacteria, are included in the schema depicted at the end of the preceding chapter. *Escherichia coli* and *Klebsiella pneumoniae* constitute normal intestinal microbiota. Species of *Salmonella*, *Shigella* and *Yersinia* are also called enteric bacteria, but they do not represent normal intestinal microbiota, although they can survive in the intestines and that is where they lodge in the case of infection. Clinically significant members of the family enterobacteriaceae and some of their physiological differences are depicted in Table 9.2.

Clinical Significance

Escherichia coli

Escherichia coli and most other fecal coliforms are most abundant in sewage and sewage-contaminated water. Therefore, the infections are often acquired by ingestion of food and beverages contaminated with untreated water. Since this is the most abundant bacterium in feces, poor personal hygiene also plays important role in the dissemination of *E. coli* related infections. *Escherichia coli* are the only clinically significant species of the genus *Escherichia*. It has been incriminated in the cases of diarrhea, gastroenteritis, septicemia, and neonatal meningitis. It is also responsible for nearly 70% of all the cases of urinary tract infections (UTI).

Virulence Factors   Many strains of *E. coli* produce endotoxins, heat labile enterotoxins (LT) and heat stable enterotoxins (ST). These toxins are usually produced by strains in O groups, including 06, 08, 015, 025, 063, 091, 0111, 0145, 0147, 0152, 0157, and 0159. Other virulence factors include pili and fimbriae which help with attachment, hemolysin, and a Shiga toxin-like substance. In addition to O antigens, two other antigens described as H and K antigens are produced by some strains.

Table 9.2 Some Physiological Characteristics of Important Members of the Family Enterobacteriaea

Species	Indole test	Urea hydrolysis	Lactose*	Mannitol*	Adonitol*	Maltose*	Glycerol*
Escherichia coli	Positive	Negative	Positive	Positive	Negative	Positive	Positive
Enterobacter cloaceae	Negative	Variable	Positive	Positive	Negative	Positive	Variable
Klebsiella pneumoniae	Negative	Positive	Positive	Positive	Positive	Positive	Positive
Proteus mirabiis	Negative	Positive	Negative	Negative	Negative	Negative	Negative
Salmonella typhi	Negative	Negative	Negative	Negative	Negative	Positive	Variable
Serratia marcescens	Negative	Negative	Negative	Positive	Variable	Positive	Positive
Yersinia pestis	Negative	Negative	Variable	Positive	Negative	Variable	Negative

* Fermentation.

Enterobacter cloacae

Enterobacter cloacae is an opportunistic pathogen in immune-compromised patients. It accounts for 4%–10% of all cases of bacteremia caused by Gram-negative bacteria. The infection is characterized by fever and may be fatal if not controlled in a timely manner. Other clinical conditions may include endocarditis, ventriculitis, meningitis, and occasionally urinary tract infections. *Enterobacter cloacae* along with *Klebsiella pneumoniae* and *Serratia marcescens* are often associated with the contamination of intravenous fluids, and are thus a major source of hospital-acquired infection. Another *Enterobacter* spp., *E. aerogenes*, is also often isolated from clinical specimens. Members of the genus *Enterobacter* generally yield a positive result in ornithine decarboxylase tests and are capable of growing in potassium cyanide. In addition, strains belonging to most species of this genus are motile.

Klebsiella pneumoniae

Klebsiella pneumoniae is the only pathogenic species of the genus *Klebsiella*. Like *E. coli*, it is a common fecal bacterium and abundantly present in sewage and untreated water. Infection can be endogenous or exogenous, and is a major cause of hospital-acquired infections. It is an important causal agent of pneumonia, lung abscess, septicemia, meningitis, and otitis externa, and a major cause of UTI. *Klebsiella pneumoniae* strains are nonmotile and yield a negative result in ornithine decarboxylase and indole tests. Most strains of *K. pneumoniae* can grow at 45°C.

Therapeutic measures may depend on the clinical manifestation. For example, oral antibiotics are frequently used for the management of UTI and intravenous antibiotics are often used to treat pneumonia. Resistance factors carried by some organisms, such as extended spectrum beta lactamases, make the use of antimicrobial resistance testing crucial in all serious infections.

Virulence Factors   Virulence factors include cell wall receptors, polysaccharide capsules, a large plasmid, enterotoxins, and K antigens.

Proteus mirabilis

Proteus mirabilis (more frequently) and *P. vulgaris* (less commonly) are important pathogenic species of the genus *Proteus*. This bacterium is often associated with UTI. *Proteus* is also occasionally incriminated in pneumonia and septicemia. The virulence factors include production of urease, which breaks down urea into ammonia, the latter being highly toxic to the cells. This also contributes to struvite deposition in the upper urinary tract. Many strains produce bacteriocin. Two other species of the genus *Proteus* are commonly encountered in clinical specimens. These include *P. penneri* and *P. vulgaris*. They can be differentiated from each other by indole, ornithine decarboxylase, esculin hydrolysis, and salicin fermentation tests. Only *P. vulgaris* strains yield a positive indole test and only *P. mirabilis* are ornithine decarboxylase positive. Most strains (but not all) of *P. penneri*, especially biotype 2, are positive in salicin fermentation and esculin hydrolysis tests.

Salmonella *Species*

Salmonella typhi has been reported mostly from humans. It is the well-known causal agent of typhoid, the deadly disease that affects millions worldwide. In the United States, approximately 400 cases are reported each year. The symptoms of typhoid may include fever, headache, gastrointestinal discomfort, loss of electrolytes, and dehydration. The disease is often fatal if not treated in a timely manner. The infection is usually food and waterborne. Chlorination of water has greatly reduced incidence of typhoid in industrialized countries, but the disease still remains a major public health concern in poor countries. Asymptomatic carriers of *S. typhi* are not uncommon.

 Salmonella paratyphi and *S. choleraesuis*, the other two pathogenic species of *Salmonella*, are commonly associated with poultry, cattle, pig, sheep, lizards, and so on. Human infections are usually acquired due to ingestion of contaminated meat, poultry, or other food products. The symptoms may include diarrhea, dysentery, bronchopneumonia, pyelonephritis, and meningitis.

Virulence Factors   The virulence factors include ability of the bacterium to adhere to and penetrate epithelial cells and its ability to multiply within the phagocytic cells. O antigens and LPS are also important virulence-related properties.

Serratia *Species*

Genus *Serratia* has several species of which seven are generally encountered in clinical specimens. These include *S. marcescens, S. ficaria, S. liquefaciens, S. odorifera, S. plymuthica, S. rubidaea,* and *S. fonticola.* Of these, *S. marcescens* is perhaps the most important. This species is characterized by positive DNAse, lipase, gelatin hydrolysis, and sorbitol fermentation, and negative arabinose fermentation tests. An interesting feature of certain *Serratia* spp. is their ability to produce characteristic red pigment at ambient temperature (20–25°C), but not at 35°C. *Serratia marcescens* is associated, although less commonly than other organisms, with UTI, bacteremia (especially in the presence of an IV catheter), and neonatal meningitis (often nosocomially acquired infections).

Shigella *Species*

Of all *Shigella* spp., *S. dysenteriae* is the most important pathogen. It causes shigellosis or bacillary dysentery. Symptoms may include fever, cramps, and diarrhea. The diarrhea may be accompanied by mucous and may eventually become bloody. Death may occur due to electrolyte imbalance. Fecal contamination of food and water is a common source of infection. Globally, shigellosis is responsible for a high degree of mortality and morbidity each year. Occasionally, *Shigella* strains are mistaken for *E. coli*. Normally, *E. coli* strains are motile, produce gas from glucose, and ferment lactose and mucate. In contrast, *Shigella* spp. are nonmotile and negative in gas production from glucose, and lactose and mucate fermentation tests.

Virulence Factors   The bacterium penetrates the intestine and invades intestinal cells, resulting in the lysis of the cells. Virulence factors include production of the deadly Shiga toxin, which is cytotoxic, enterotoxic, and neurotoxic.

Yersinia *Species*

Yersinia pestis causes plague, the deadliest of all infectious diseases, which has left a vivid mark in the history of medieval Europe. During the recent years, authentic cases of plague have been reported from many parts of the world including the United States and India. Plague can be divided into two distinct clinical forms:

1. Bubonic plague is acquired via fleas that infect rodents, such as rats. The symptoms include fever and lymphadenitis.
2. Pneumonic plague is transmitted from infected persons to healthy ones. The symptoms include pneumonia. This stage is a highly contagious.

Virulence Factors   Virulence factors include the ability of the bacterium to invade tissue, its ability to survive in the phagosomes, and virulence plasmids. Several strains are known to produce pesticin, which is a bacteriocin. Bacteriocins are toxic proteins that inhibit growth of closely related bacteria, and with certain reservations, they can be called narrow spectrum antibiotics.

Laboratory Diagnosis   Clinically significant species of *Yersinia* grow on most media that have been discussed earlier. Storage at 4°C over an extended period can help with the enrichment and enhance isolation in culture. But this technique of cold storage enrichment has very little diagnostic value. Gram negative rods can be easily seen in the stained smears (Fig. 9.1). Serological test for *Y. enterolytica* O8 also has only a limited value. The diagnosis of plague is generally made on the basis of clinical findings with limited support from the laboratory observations. Of the several species of genus *Yersinia* (10 species by some counts), *Y. enterocolitica, Y. pseudotuberculosis, Y. kristensensii, Y. intermedia*, and *Y. frederiksenii* can be also encountered in clinical specimens. Therefore, species differentiation is required. *Yersinia pestis* strains are generally nonmotile, indole negative, do not hydrolyze urea, and yield a positive ornithine decarboxylase test.

Antibiotic Sensitivity   Susceptibility of enteric bacteria to antimicrobial agents is highly variable and it may differ from strain to strain. Resistance to antibiotics that were previously considered effective is quite common these days. A sensitivity test is, therefore, highly desirable. Usually, the local health care institutions keep a track of their susceptibility pattern. The antibiotics that may be effective include ticaracillin, clavulanic acid, second-generation cephalosporins, fluoroquinolones, kanamycin, tetracycline, and trimethoprim sulfamethoxazole.

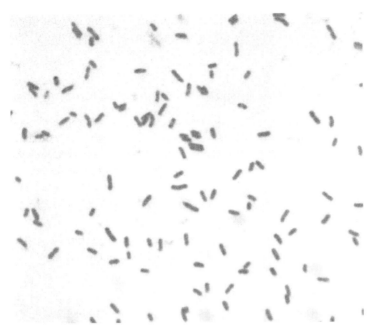

Figure 9.1.   A Gram-stained smear prepared from a *Y. pestis* strain isolated from a case of plague (courtesy: CDC).

Citrobacter *Species*

Citrobacter spp. are motile bacilli that may or may not always ferment lactose. Another distinguishing feature is their negative reaction in lysine decarboxylase tests. In addition to differences shown in the Table 9.3, the three species discussed here can be differentiated from each other on the basis of indole test and hydrogen sulfide production on triple sugar agar. Infections caused by *Citrobacter* spp. are often nosocomial and usually noted in immunocompromised persons. Species frequently isolated from human cases include *C. freundii, C. diversus*, and *C. amalonaticus*. These often occur in the urinary tract, the respiratory tract, wounds, and cutaneous infections. Cases of septicemia, brain abscess, and meningitis have been occasionally noted in association with *C. freundii* and *C. diversus*. Strains of *Citrobacter* spp. have traditionally been sensitive to ampicillin and cephalothin, but instances of resistance to these drugs is on rise.

Edwardsiella *Species*

Two species of the genus *Edwardsiella, E. tarda* and *E. hoshinae*, are clinically significant. These species are not uncommon in human environments and are occasionally isolated from reptiles, fishes, and other animals. These two species are associated with gastrointestinal disorders mostly in the tropics. *Edwardsiella tarda* has occasionally been held as the causal agent in cases of septic shock, hepatic abscess, meningitis,

Table 9.3   Some Physiological Differences among Certain Members of the Family
Enterobacteriaceae That Are Relatively Uncommon Pathogens

Taxa	Urea hydrolysis	Lactose fermentation	Adonitol fermentation	Maltose fermentation	Glycerol fermentation
Kluyvera ascorbata	Negative	Positive	Negative	Positive	Variable
Morganella morganii	Positive	Negative	Negative	Negative	Negative
Providencia stuartii	Negative	Negative	Negative	Negative	Variable
Proteus rettgeri	Positive	Negative	Positive	Negative	Variable
Edwardsiella tarda	Negative	Negative	Negative	Positive	Negative
Edwardsiella hoshinae	Negative	Negative	Negative	Positive	Variable
Citrobacter freundii	Positive	Variable	Negative	Positive	Positive
Citrobacter diversus	Positive	Variable	Positive	Positive	Positive
Citrobacter amalonaticus	Positive	Variable	Negative	Positive	Negative

and wound infection. In addition to the differences summarized in Table 9.3, the two
species can be differentiated on the basis of indole test and hydrogen sulfide production on triple sugar agar, which are positive in the case of *E. tarda* and negative for
E. hoshinae. *Edwardsiella* strains are generally sensitive to ampicillin, cephalosporins, fluoroquinolones, and trimethoprim-sulfamethoxazole.

Morganella morganii

Morganella morganii is an opportunistic pathogen, mostly noted in the elderly and in
hospitalized patients recovering from major surgeries. More recently, it has been
noted, rather often, in the cases of bacteremia. Strains of *M. morganii* may be occasionally mistaken for *Proteus* and *Providencia* spp. Unlike *Proteus* spp., *M. morganii*
strains are hydrogen sulfide negative. They differ from *Providencia* spp. with reference to citrate utilization tests, which are negative for *M. morganii* and positive for
Providencia spp. Strains of *M. morganii* are sensitive to ampicillin and cephalothin.

Kluyvera ascorbata

Strains identified as *K. ascorbata* have been mostly isolated from fecal matter, urine
samples, and sputum. However, the reported cases in which this bacterium is believed

to be the causal agent are limited to urinary tract infections and peritonitis. Strains of *K. ascorbata* can be mistaken for *E. coli*. They generally yield positive results in citrate utilization, lysine decarboxylation, and sorbitol and raffinose fermentation tests while *E. coli* are mostly negative. *Kluyvera ascorbata* strains are sensitive to ampicillin and cephalothin.

Providencia *Species*

From clinical perspective, the most important species of *Providencia* are *P. stuartii* and *P. rettgeri*. The infections are mostly noted in elderly persons with urinary tract disorders. These species are also known to colonize skin damaged by severe burns. Urinary tract infections are more common in patients with catheters. Strains of *Providencia* spp. are generally negative in hydrogen sulfide tests. They can be mistaken for *Shigella* spp., but can be differentiated on the basis of motility and citrate utilization tests, which are generally positive for *Providencia* spp. and negative for *Shigella* spp. Most strains of *P. rettgeri* and *P. stuartii* are sensitive to nitrofurantoin and beta lactam antibiotics. Important physiological differences among the species of *Providencia* and other uncommon pathogens belonging to the family Enterobacteriaceae are summarized in Table 9.3.

GLUCOSE NONFERMENTERS

Glucose nonfermenters, also known as NFB (nonfermenting bacteria), mostly include common environmental bacteria that seem to prefer wet conditions. Some NFB are also important opportunistic pathogens. These include the following:

- *Acinetobacter*
- *Alcaligenes*
- *Flavobacterium*
- *Moraxella*
- *Pseudomonas*
- *Xanthomonas*
- *Burkholderia*

Growth Requirements

NFB grow well on blood agar and on some of the other media used for Enterobacteriaceae, especially MacConkey agar. Selective media for *Pseudomonas aeruginosa* include acitamide agar, cetrimide agar, and OFPBL medium for *Burkholderia* (*Pseudomonas*) *cepacia*.

Except for a few exceptions, a majority of NFBs are strict aerobes, fast growers, and prefer incubation at 35°C. Distinguishing colony characteristics of some NFBs on blood agar are described below, along with other relevant features.

Acinetobacter Species

On blood agar, colonies are raised, creamy, circular, and opaque, with hemolysis. It also grows well on MacConkey agar. Members of this genus are commonly present in soil and water. They can also be isolated from skin. This bacterium is not known to infect humans with unimpaired immunity. However, its association with nosocomial infections, pneumonia, meningitis, UTIs, septicemia, and wound infection has been reported. *Acinetobacter baumannii* is considered pathogenic to humans. Lately, this bacterium has become a more common cause of infections associated with stays in the intensive care unit and prolonged hospitalizations.

Alcaligenes Species

A species of clinical significance is *A. faecalis*. These are oxidase positive and grow well on MacConkey agar. On blood agar, the colonies are nonpigmented, glistening, convex, and produce a fruity odor. Like *Acinetobacter baumannii*, *Alcaligenes faecalis* is mostly associated with infections in immune-compromised patients. Its isolation from sputum and urine has been occasionally reported. Also, like *A. baumannii*, it is known to cause nosocomial septicemia in some cases.

Flavobacterium Species

Flavobacterium meningosepticum is the only species of clinical significance. It is oxidase positive and it may or may not grow on MacConkey agar. On blood agar, the colonies are circular, convex, smooth, and yellowish orange. This bacterium has been incriminated in a few cases of neonatal meningitis. Cases of meningitis, pneumonia, and septicemia in adults have been reported on rare occasions. Other notable species are *F. indologenes, F. odoratum,* and *F. thalpohilum,* which have been noted in association with the rare cases of meningitis, septicemia, urogenital tract disorders, and wounds. But their role in the causation of the disease is uncertain.

Moraxella Species

Important species of the genus *Moraxella* include *M. atlantae, M. lacunata, M. nonliquefaciens,* and *M. osloensis*. These species are oxidase positive and they may or may not grow on MacConkey agar. On blood agar, the colonies are tiny and translucent. *Moraxella* spp. are normally present on mucous membrane, but they are generally considered harmless. However, strains identified as *M. lacunata* are occasionally associated with conjunctivitis and corneal infections. *Moraxella osloensis* are generally present in the genitourinary tract.

Xanthomonas Species

Xanthomonas maltophilia, formerly known as *Pseudomonas maltophilia*, is a clinically significant species of the genus *Xanthomonas*. Most strains are oxidase negative and grow well on MacConkey agar. On blood agar, the colonies are lavender colored at first and turn grayish green with age. Some strains produce an ammonia-like odor. Xanthomonas maltophilia strains are sometimes isolated from clinical specimens in hospitalized patients and involved in cases of pneumonia, bacteremia, endocarditis, meningitis, and UTI. Most infections are nosocomial and often noted in immunocompromised subjects. In a majority of such cases, infections are associated with the instrumentation or manipulative procedures.

Burkholderia Species

An important species of the genus *Burkholderia,* from the clinical perspective, is *B. cepacia*, formerly known as *Pseudomonas cepacia*. Like *P. aeruginosa, B. cepacia* is also widely distributed in nature. Most strains are oxidase negative and grow on a wide range of media including MacConkey agar and blood agar. Selective media have been developed for the isolation of this bacterium. It is one of the most important causal agents of respiratory tract infections in cystic fibrosis patients and an occasional cause of nosocomial infections. Urinary tract infections are often associated with the use of the catheter. This bacterium has also been isolated from intravenous catheters. Septicemia associated with the use of contaminated catheters has been reported. *Burkholderia cepacia* has also been associated with the nosocomial outbreaks of pneumonia and septicemia. Like *P. aeruginosa, B. cepacia* infections are difficult to treat. Antimicrobial resistance testing is important in determining the most ideal therapy.

Pseudomonas Species

Genus *Pseudomonas* has many species, but the most important member is *P. aeruginosa*. A majority of strains are oxidase positive and they grow well on MacConkey agar. On blood agar, the colonies are usually large, irregular, round, and have a ground glass appearance with grape-like odor. *Pseudomonas* spp. are dynamic bacteria, abundantly present in aquatic systems, and play an extremely important role in recycling the biomass in ecosystem.

In this era of indiscriminate use of antibiotics, corticosteroids, and immune suppressants, it is hard to draw a sharp line between pathogenic and nonpathogenic microorganisms. However, of all the bacteria listed as glucose nonfermenters, *P. aeruginosa* is of immense clinical importance. It is an important opportunistic pathogen and is also frequently incriminated in nosocomial infections such as pneumonia and bacteremia. Of the other notable species of this genus, *P. fluorescens* and *P. putida* are normally present in the upper respiratory tract and are occasionally isolated from blood, cerebrospinal fluid, pleural fluid, urine, and wounds, but their actual role in the causation of the diseases is uncertain. Another species, *P. mallei*, has been noted in association with glanders, a disease often noted in horses. *Pseudomonas pseudomallei*, a closely related species, is generally associated with glander-like disease in humans, mostly in Southeast Asia. But this bacterium is normally present in soil and water in many parts of the world. Perhaps next to *P. aeruginosa*, *P. stutzeri* is of some clinical significance. It has been isolated from blood, cerebrospinal fluid, discharges from the middle ear, sputum, and urine and is believed to be the causative agent at least in some of the cases. Among the other notables are *P. alcaligenes, P. gladioli,* and *P. pickettii,* which have been isolated from blood, urine, cerebrospinal fluid, and the respiratory tract of patients suffering from empyema, endocarditis, bacteremia, meningitis, and cystic fibrosis.

Disease

Pseudomonas aeruginosa is known to cause a wide range of clinical complications including burn and wound infection, pneumonia, chronic pulmonary disorders, septicemia, otitis externa, and UTI. *Pseudomonas aeruginosa* is also known to cause major complications in cystic fibrosis cases. Diseases caused by *P. aeruginosa* are generally rare in immunocompetent persons. Infections are mostly noted in patients with underlying conditions, such as intravenous drug use, serious burn injuries, granulocytopenia, long-term antibiotic therapy, and indwelling catheter or other devices.

Virulence Factors   Virulence factors include endotoxin (LPS), exotoxins A and S, cytotoxins (leukocidins), proteinases, phospholipases (responsible for the destruction of pulmonary surfactants), and pili that help with adherence. Extracellular proteinases are believed to be responsible for the degradation of a wide range of proteins associated with structural components and homeostasis.

Antibiotic Sensitivity   *Pseudomonas aeruginosa* is resistant to many of the commonly used antimicrobial agents. Therefore, sensitivity testing is a necessity. Some effective agents include aminoglycosides, ceftazidine, piperacillin, carbapenems, and ciprofloxacin.

UNCOMMON NONFERMENTATIVE TAXA

Chryseomonas luteola

Strains of *C. luteola* are characterized by the presence of polar flagella. They are motile, oxidase negative, and grow well on MacConkey agar. Isolates identified as *C. luteola* have been occasionally noted in the cases of wounds and abscesses, and have also been associated with a few cases of peritonitis and bacteremia, mostly in critically ill persons.

Eikenella corrodens

Eikenella corrodens strains are oxidase positive and nonmotile. They do not grow on MacConkey agar. This bacterium is normally present on the mucous membrane, mostly in the nasopharynx and gastrointestinal tract. Strains have been also isolated from cases of meningitis, and infections involving the neck and head, as well as bite wounds, mostly in immunocompromised patients.

Kingella kingae

Kingella spp. are facultative anaerobic, coccobacilli that are nonmotile, and oxidase negative, and may or may not grow on MacConkey agar. Strains of *K. kingae* have been occasionally isolated from blood, urine, throat, and wound samples. It is an opportunistic pathogen, and has been occasionally incriminated in the cases of endocarditis, arthritis, osteomyelitis, and septicemia. A majority of patients in such cases were children.

Weeksella virosa

Weeksella virosa strains are also nonmotile and oxidase positive. Their growth on MacConkey agar is variable, mostly poor. They inhabit the urogenital tract of approximately 2% of women and are frequently isolated from the vaginal swabs of women who have had multiple sexual partners. There seems to be a relationship between the number of sexual partners and frequency of isolation of this bacterium from vaginal swabs. It has been occasionally incriminated in the cases of urethritis, peritonitis, pneumonia, sepsis, and vaginal infection.

Antibiotic Sensitivity

Antibiotic sensitivity of these abovementioned nonfermentative Gram-negative bacilli can be hard to predict. Therefore, it is important to perform sensitivity tests, if their role in the causation of disease is suspected.

Chapter 10

Miscellaneous Gram-Negative Bacteria

This chapter covers several Gram-negative but unrelated taxa. We did not mean to make this chapter a dumping ground, but they are all included in this chapter simply for the sake of brevity. The important taxa include the following:

- *Brucella melitensis*
- *Bordetella pertussis*
- *Francisella* spp.
- *Pasteurella* spp.
- *Vibrio cholerae*
- *Campylobacter* spp.
- *Helicobacter* spp.
- *Legionella* spp.
- *Gardnerella vaginalis*
- *Chlamydia* spp.
- *Rickettsia rickettsii*

BRUCELLA MELITENSIS

Brucella spp. are Gram-negative, aerobic coccobacilli. Earlier, the genus *Brucella* was divided into six species but subsequent genetic studies have established that all the six species essentially represent different variants of *Brucella melitensis*. *Brucella melitensis* is an important intracellular pathogen commonly associated with cattle, swine, dogs, and rats, which can also serve as the carriers. Goats are believed to be the natural reservoir for *B. melitensis*. Goats themselves seldom suffer from *Brucella* infections, but pass on the bacterium in their milk and subsequently in

A Concise Manual of Pathogenic Microbiology, First Edition. Saroj K. Mishra and Dipti Agrawal.
© 2013 Wiley-Blackwell. Published 2013 by John Wiley & Sons, Inc.

cheese. Human cases of brucellosis are not uncommon, particularly in poor Asian and African countries. The infections are often acquired by consumption of *B. melitensis*-contaminated goat milk and cheese.

Disease

Brucella melitensis causes brucellosis, which is characterized by fever, enlarged spleen, weight loss, and arthritis. The fever shows a high degree of fluctuation with phases when the body temperature is high often for weeks followed by days with a low grade or no fever. Another species, *B. abortus* (more correctly *B. melitensis biotype abortus*), causes abortion in the cattle.

Laboratory Diagnosis

Samples of blood or any other clinical specimen should be inoculated into a brain heart infusion agar or trypticase-soy broth. Cultures are incubated in a CO_2 incubator for 2–7 days. Subcultures on blood or chocolate agar plates are frequently needed. Selective media and numerous serological tests are also available. Visible translucent colonies develop after 72 hours and turn gray with time. Microscopically, *B. melitensis* are Gram-negative coccobacilli (Fig. 10.1). In histopathological preparations, they are mostly intracellular. A serum agglutination test is also available and useful.

Figure 10.1.   A Gram-stained smear showing coccobacilli of *Brucella melitensis*.

Antibiotic Sensitivity

A combination therapy involving rifampin and doxycycline is preferred. Other effective agents include streptomycin and trimethoprim sulfa.

BORDETELLA PERTUSSIS

Bordetella pertussis is an important pathogen of worldwide occurrence. Annually, more than 50 million persons are known to suffer from *B. pertussis* infections. This bacterium is found only in humans. Its isolation from other animals or environmental sources has not been reported. This bacterium was first isolated by Bordet and Gengou from a case by whooping cough and later named after Bordet in his honor. The genus has only two species. A second species, *B. avium*, a causal agent of coryza (rhinotracheitis) in turkeys was added in 1984. *Bordetella* spp. are aerobic, chemoorganotrophic, nonmotile, flagellated coccobacilli.

Disease

Bordetella pertussis causes pertussis (whooping cough), which is a highly contagious bioaerosol-borne disease that frequently affects children. Initially, the bacterium attaches to the cells of the respiratory epithelium, causing progressive irritation of the upper trachea. The resulting cough can be so severe that the patient may find it hard to breathe. Death may occur due to airway obstruction. Usually a childhood infection gives lifelong immunity but infections in elderly patients who have had childhood whooping cough is not uncommon. However, such cases only have a persistent cough devoid of whooping.

The disease is well known to cause high morbidity and mortality, especially in developing countries. Prior to the widespread use of vaccine, pertussis was quite common in United Stated and other industrialized countries. Currently, however, it is being increasingly recognized as a pathogen in adults with waning immunity.

Virulence Factors

Bordetella pertussis produces an exotoxin called pertussis toxin. Pili help with the attachment. In addition, two other toxins, dermonecrotic toxin and tracheal cytotoxin, have been identified. The tracheal cytotoxin is believed to kill ciliated cells in the upper respiratory tract area and stimulates release of interleukin-1, which probably plays a role in the fever. The role of dermonecrotic toxin is uncertain at present.

Laboratory Diagnosis

Nasal and throat swabs should be streaked on special media, such as Bordet-Gengou medium. Incubation is done at 35°C, aerobically, or in the presence of 3%–8% CO_2.

Several serological tests and molecular genetics-based techniques are also available for rapid diagnosis and species identification. *Bordetella* pertussis strains are oxidase positive and urease negative, aerobic, Gram-negative coccobacilli.

Antibiotic Sensitivity

Bordetella pertussis is sensitive to erythromycin. Excellent preventive vaccines are available and widely used in the industrialized countries.

FRANCISELLA TULARENSIS

Like *Bordetella* and *Brucella* spp., *Francisella* are aerobic, nonfermentative coccobacilli that are almost always associated with an infection. The genus *Francisella* is believed to have two species, *F. tularensis* and *F. philomiragia*. Of the two, *F. tularensis* is of a greater clinical significance.

Disease

Francisella tularensis causes tularemia, also known as rabbit fever, tick fever, and glandular fever. The disease is considered zoonotic and mostly results from exposure to rodents, rabbits, or ticks. The infection results from the handling of infected animals, or following a tick bite. The incubation period is generally 3–5 days. The symptoms develop rather abruptly and may include fever, chill, and malaise. In some cases, painful lesions develop on the skin near the enlarged lymph nodes, leading to lymphadenopathy and bacteremia. Another clinical manifestation of tularemia, called oculoglandular tularemia, develops following a direct infection of the eyes by exposure to the bioaerosols from infected patients or through contaminated fingers. Infected individuals develop painful conjunctivitis and regional lymphadenopathy. Pneumonic tularemia, yet another type of clinical manifestation, may result from the inhalation of infectious bioaerosol and can be fatal if not treated in a timely manner.

Virulence Factors

Not much is known about the virulence factors associated with *F. tularensis* except for the presence of antiphagocytic capsules. By virtue of being an intracellular pathogen, the bacterium resists complement-mediated killing and phagocytosis.

Laboratory Diagnosis

Clinical materials, such as exudates from the lesions, biopsied lymph nodes, and respiratory secretion, should be collected with utmost care. Because *F. tularensis* is a highly contagious pathogen, proper safety precautions must be observed when collecting or processing clinical specimens. The bacterium can be isolated on glucose-cysteine blood (GCB) agar supplemented with thiamine. *Francisella tularensis* can also be isolated, albeit less successfully, on charcoal-yeast extract agar and Thayer-Martin medium. The incubation is done aerobically at 35°C for 3–5 days. On GCB agar, small colonies measuring about 3 mm develop after 3–4 days.

Antibiotic Sensitivity

Cases of tularemia can be treated with streptomycin, chloramphenicol, or tetracycline. However, streptomycin is generally considered the drug of choice. Most strains of *F. tularensis* produce beta lactamases; therefore, penicillins are not indicated.

PASTEURELLA SPECIES

Similar to *Brucella, Bordetella*, and *Francisella* spp., members of the genus *Pasteurella* are also Gram-negative aerobic coccobacilli. However, unlike the other three, members of the genus *Pasteurella* are fermentative. Most strains of *Pasteurella* spp. grow slightly faster than those belonging to the other three genera. Clinically, important species include *P. multocida, P. pneumotropica, P haemolytica*, and *P. ureae* (now *Actinobacillus ureae*). Based on the results of genetic studies, currently a process to revisit genus *Pasteurella* is under way, and it is quite possible that some of the species will be merged or split to create newer species.

Species belonging to genus *Pasteurella* have been isolated from wild and domesticated animals, but most isolates of *P. ureae* have originated from human sources only. A majority of strains of *P. ureae* have been isolated from the respiratory tract of apparently healthy persons, but it has been occasionally incriminated in the cases of septicemia and meningitis. *Pasteurella multocida*, which is normally present in the respiratory tract of numerous wild and domestic animals, has been occasionally implicated in cellulitis-like lesions or abscesses developing at the site of an animal bite. The infection is known to progress into osteomyelitis and arthritis. In rare instances, *P. multocida* has been noted in association with pneumonia, emphysema, and lung abscess. *Pasteurella ureae*, though frequently isolated from human sources, has been only rarely found in association with septicemia and meningitis in humans.

Virulence Factors

Members of the genus *Pasteurella* are encapsulate and that certainly interferes with phagocytosis. Also, several strains produce soluble toxins which could play a role in its pathogenesis.

Laboratory Diagnosis

Pasteurella spp. grow well on blood agar and chocolate agar. Cultures are generally incubated aerobically at 35°C. Visible colonies develop after 24 hours. The colonies are generally smooth with a grayish appearance. As stated above, *Pasteurella* spp. are nonmotile, Gram-negative bacilli that are fermentative, and mostly oxidase and catalase positive.

Antibiotic Sensitivity

Pasteurella spp. are generally sensitive to penicillin and possibly to tetracycline and cephalosporins. They are mostly resistant to aminoglycosides.

VIBRIO CHOLERAE

Members of the genus *Vibrio* are abundantly present in the fresh and marine waters. It is a large genus that is generally divided into 26 species, 12 of which are believed to be clinically significant. Of these, *V. cholerae*, being the causal agent of cholera, is perhaps the most important and best known. Other species, such as *V. mimicus, V. parahaemolyticus, V. alginolyticus, V. fluvialis,* and *V. furnissii* are mostly implicated in gastroenteritis, and occasionally in the cases of ear and wound infections, conjunctivitis, and other complications.

Like other members of the genus *Vibrio, V. cholerae* is primarily an aquatic bacterium and is frequently found in the sewage contaminated water. It is a facultative anaerobe, Gram-negative curved rod with a polar flagellum (Fig. 10.2). Tests

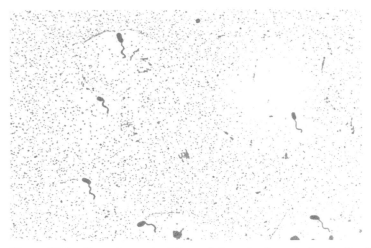

Figure 10.2.   Photomicrograph of *Vibrio cholerae* showing a single polar flagellum in a digitally colorized Leifson flagella stain (source: CDC). See color insert.

for indole are positive, urease negative, and oxidase positive, and most strains reduce nitrate to nitrite.

Disease

Vibrio cholera is the principal causal agent of the deadly disease cholera, responsible for millions of deaths and debilitating morbidity each year in most of the developing countries. The affected areas include much of Asia, Africa, and Latin America. During the period 1991–1995, more than a million people suffered from cholera in Latin America alone, resulting in the deaths of nearly 10,000 persons. No clear data are available from the endemic areas of Asia and Africa, but the extent of mortality and morbidity is believed to be quite high. Infection is usually acquired by ingestion of food or beverages contaminated with the fecal matter. The incubation period is often less than 3 days. Symptoms include diarrhea and vomiting that quickly lead to severe loss of fluid and electrolytes. More than 70% of untreated patients die. Among the most important biotypes are *V. cholerae* O1 and *V. cholerae* O139. Other species of genus *Vibrio* that are often indicated in human diseases, such as *V. parahaemolyticus, V. alginolyticus, V. mimicus* and several others, will not be discussed here.

Virulence Factor

Cholera toxin is commonly noted in the strains of *V. cholera biotype* O1. Other virulence factors include pili, adhesion factors, siderophores, and neuraminidase. The strains of *V. cholerae* other than biotype O1 produce heat stable enterotoxins, heat labile hemolysin, and proteinases. Other species are also known to produce enterotoxins and hemolysin.

Laboratory Diagnosis

Vibrio cholera is not a fastidious bacterium. Thiosulfate–citrate–bile salt sucrose (TCBS) agar is an excellent isolation medium. However, not all species of *Vibrio* grow on this medium. Incubation in enrichment broth (alkaline peptone water) may be required. The cultures can be incubated aerobically or anaerobically at 35°C for 24 hours. On TCBS medium, colonies of *V. cholerae* are yellow and measure about 2 mm in diameter. Colonies of *V. mimicus* and *V. parahaemolyticus* that do not ferment sucrose develop a green color. *Vibrio* spp. generally grow well on sheep blood agar and develop a greenish color. Serologically, *V. cholerae* O1 are believed to have three antigenic factors (A, B, and C). Gene probes are also available for specific identification.

Antibiotic Sensitivity

Fluid and electrolyte balance should be restored promptly. Antibiotics help to decrease the duration of diarrhea by 50% and decrease the excretion of the affected person by 1 day. Tetracylcines have been used in most cases to treat *V. cholera* infections. However, some areas have strains that are resistant to tetracyclines. In such cases, macrolides and fluoroquinolones are acceptable alternatives. Vaccines that provide protection for a limited period are also available.

AEROMONAS SPECIES

Members of the genus *Aeromonas* are a group of dynamic bacteria in the sense that they can grow at pH 4.5 to 9.0 and at temperatures ranging from 10 to 45°C. Six species, including *A. hydrophila, A. caviae, A. veronii, A. jandaei, A. schubertii*, and *A. trota*, are considered medically significant. *Aeromonas* spp. cause gastroenteritis characterized by watery diarrhea, often accompanied by fever, nausea, and abdominal pain. The infections are usually food- and waterborne. These bacteria have also been indicated in cases of septicemia, respiratory and urinary tract infections, peritonitis, and conjunctivitis. *Aeromonas* infections appear to be more common in Southeast Asia. No specific virulence factor has been identified except that many strains produce cytotoxins and enterotoxins.

Laboratory Diagnosis

Aeromonas spp. grow well on blood agar, MacConkey agar, and Hektoen agar. Selective media are also available for their isolation from feces and other heavily contaminated specimens. Certain specimens may require enrichment, which can be achieved by incubating the sample in alkaline peptone water or Gram-negative broth. *Aeromonas* spp. are facultative anaerobes. Visible colonies appear within 24 hours. Except for *A. caviae* strains, most species cause a strong beta hemolysis on blood agar.

Antibiotic Sensitivity

Aeromonas spp. produce a variety of beta lactamases and are generally resistant to penicillins and first-generation cephalosporins. Antibiotic sensitivity tests are strongly recommended.

CAMPYLOBACTER SPECIES

Campylobacter are Gram-negative, microaerophilic, aerobic, nonspore forming curved rods that are motile and have one polar flagellum (Fig. 10.3). They are oxidase positive. The main pathogenic species are *C. jejuni* and *C. pylori*. *Campylobacter pylori* is also referred to as *Helicobacter pylori* by some microbiologists. *Campylobacter* is similar to *Vibrio* in appearance, but they are nonfermentative and their DNA base ratio is lower (29% compared with 45% in *Vibrio*). The natural reservoirs of *C. jejuni* include poultry, pigs, and cattle.

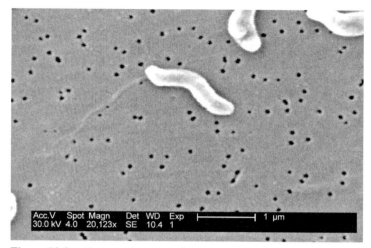

Figure 10.3.   Electron micrograph of *C. jejuni* (note the curved rod with a polar flagellum).

Table 10.1   Major Similarities and Differences between *Campylobacter jejuni* and *C. pylori* Strains

Properties (tests)	*C. jejuni*	*C. pylori* (*H. pylori*)
Oxidase	Positive	Positive
Catalase	Positive	Positive
Nitrate reduction	Positive	Negative
Sensitivity to cephalosporins	Resistant	Sensitive

Disease

Campylobacter jejuni is the leading cause of enteritis all over the world. Its incubation period is about 2–10 days. Symptoms include diarrhea with bloody stool, abdominal pain, and fever. *Campylobacter pylori* is generally associated with type B gastritis (gastric ulcer or peptic ulcer). There appears to be a strong relationship between chronic gastritis and gastric carcinoma.

Laboratory Diagnosis

Suitable clinical specimens are streaked on blood agar and campy agar; the latter is a selective medium. Incubation is done at 35°C in the presence of 5%–10% CO_2. Currently, dependable campy kits are available for the rapid diagnosis. Some of the physiological differences among *C. jejuni* and *C.* (*Helicobacter*) *pylori* are summarized in Table 10.1.

Figure 10.4.   Gram-negative bacilli with some unidentified filamentous structures seen in a clinical specimen obtained from a patient suffering from Legionnaire's disease (source: CDC).

Antibiotic Sensitivity

Campylobacter are sensitive to erythromycin, but antibiotic therapy is usually not indicated in *C. jejuni* infections. Clinical management of *C. pylori* infection requires a combination of antibiotics and a proton pump inhibitor. Tetracyclines and macrolides are often used for therapy, but worldwide resistance is increasing.

LEGIONELLA SPECIES

Legionella spp. are aerobic, nonsporulating, motile, Gram-negative rods (Fig. 10.4). Unlike to other Gram-negative bacteria, but similar to *Corynebacterium* and *Mycobacterium*, *Legionella* strains contain branched fatty acid in the cytoplasmic membrane. They are widely distributed in environment and commonly present in cooling towers. There are 32 recognized species, but *L. pneumophila* is the most important pathogenic species.

Disease

Legionella pneumophila is an intracellular pathogen. The incubation period may range from 2 to 10 days. The disease variously known as Legionnaires' disease, legionellosis, or Pontiac fever may present two distinct clinical conditions:

- Pneumonia often accompanied by fever, chills, and malaise. This form of the infection can be fatal if not treated.
- Febrile illness, which can be self-limiting.

Table 10.2   Some Differentiating Characteristics of Commonly Isolated *Legionella* spp.

Species	Oxidase test	B-lactamase production	Motility	Hippurate hydrolysis	Gelatin liquification
L. pneumophila	Positive	Positive	Positive	Positive	Positive
L. feeleii	Negative	Negative	Positive	Variable	Negative
L. cherrii	Negative	Positive	Positive	Negative	Positive
L. erythra	Positive	Positive	Positive	Negative	Negative
L. adelaidensis	Negative	Negative	Positive	Negative	positive
L. israelensis	Negative	Variable	Positive	Negative	Positive

Virulence Factors

No particular virulence factor is known except that *L. pneumophila* can grow in alveolar macrophages and is capable of resisting phagolysosome fusion.

Laboratory Diagnosis

Legionella pneumophila does not grow on routine laboratory media. Throat swabs should be streaked on buffered charcoal-yeast extract agar containing alpha–ketoglutarate (BCYE-alpha). The bacterium can survive at temperatures from 4°C to 55°C. Incubation is done at 35°C for 5–7 days under aerobic condition. The colonies grow to about 3–4 mm in 7 days. They are convex and circular, grayish in color with a glistening texture and a ground glass appearance. Some of the significant differences among the commonly isolated species of *Legionella* are depicted in Table 10.2.

Antibiotic Sensitivity

Most strains of *L. pneumophila* are sensitive to fluoroquinolones and macrolides.

GARDNERELLA VAGINALIS

Gardnerella vaginalis is an endogenous opportunistic "pathogen." The bacterium is an anaerobic, fastidious, and Gram-variable bacillus.

Disease

Gardnerella vaginalis causes vaginosis, a disease characterized by thick vaginal discharge with an offensive odor. The condition must be differentiated from vaginitis, which is an inflammatory disease caused by yeast *Candida* spp. or protozoa *Trichomonas vaginalis*. Vaginosis is devoid of inflammatory symptoms, such as itching and burning sensations. The condition may be caused by a decline in the vaginal population of normal resident microbiota, especially *Lactobacillus acidophilus*.

Laboratory Diagnosis

A vaginal swab, preferably collected in the morning, immediately after the subject gets up, is the most suitable clinical specimen. It should be streaked on human blood bilayers agar with Tween 80 or Vaginalis agar containing 2% human blood, and incubated at 35°C in a CO_2 incubator for 48 hours. The colonies are β-hemolytic, opaque, convex, and grayish in color and measuring about 0.5 mm in diameter. Confirmatory tests include a positive hippurate and starch hydrolysis. Tests for α-glucosidase are positive and for the β-glucosidase are negative.

Antibiotic Sensitivity

The drug of choice is metronidazole, either orally or intravaginally, which leads to a high rate of cure. *Gardnerella vaginalis* infection can recur. Tinidazole is a second-generation nitroimidazole which can be used in these cases. *Lactobacillus* spp., used to restore normal vaginal microbiota, may also be employed. Men may serve as carriers. However, simultaneous treatment of the male partner has not been proven to be beneficial in preventing recurrence of bacterial vaginosis in female partners, though additional trials are needed.

CHLAMYDIA SPECIES

Members of the genus *Chlamydia* are very small (<1 μm in size), Gram-negative, irregular in shape, intracellular pathogens. Important pathogenic species include *C. trachomatis* and *C. pneumoniae*, which are known to infect humans only.

Disease

Chlamydia trachomatis is the well-known cause of nongonococcal urethritis, and perhaps the most common sexually transmitted disease in the industrialized world, affecting nearly 1.5 million women in the United States alone. The infection may eventually involve the bladder, kidneys, and cervix, leading to infertility. Trachoma, a chronic form of eye infection, is another common clinical condition that is noted worldwide.

   Chlamydia pneumoniae causes a milder form of pneumonia, and also infections of the pharynx or throat. Additional modes of transmission may include contact with infectious droplets, but it has not been well defined. The infection is usually mild and self-limited. Sero-prevalence rates are high by adulthood.

Lab Diagnosis

Fluorescence antibody technique, ELISA, and tissue culture using HeLa (Henrietta Latousche) cells can be useful. *Chlamydia* strains cannot be cultured on standard laboratory media.

Antibiotic Sensitivity

Most strains of *Chlamydia* spp. are sensitive to tetracyclines, quinolones, and macrolides.

RICKETTSIA RICKETTSII

Rickettsia rickettsii are Gram-negative, irregularly shaped, ultra-small (less than 0.1 µm in diameter), and obligate intracellular parasites. Their natural habitats are wood ticks (*Dermacentor andersoni*), dog ticks (*Dermacentor variabilis*), and certain other tick species in Latin America. *Rickettsia rickettsii* causes Rocky Mountain spotted fever. The term can be misleading because the disease is most common in the Eastern and Southeastern United States. A large number of cases are reported from Texas and Oklahoma.

Disease

Rocky Mountain spotted fever often begins with a rash that is present between the third and fifth day following the tick bite, and is generally accompanied by fever, nausea, and headache. Death may occur due to multiorgan failure, with the highest mortality being in the very young and the elderly. It is an important zoonotic disease.

Laboratory Diagnosis

Rocky Mountain spotted fever is usually diagnosed on the basis of clinical symptoms and case history. Serological tests are of a limited value. Since the causal agent is an obligate parasite, it cannot be cultured on laboratory media.

Antibiotic Sensitivity

Infections by *R. rickettsii* response well to tetracycline and macrolides.

BACTEROIDES SPECIES

Bacteroides spp. are anaerobic, Gram-negative bacilli that account for a large number of anaerobic bacteria isolated from human sources. They are generally present in the mouth and gastrointestinal tract, and are occasionally associated with periodontal lesions. A majority of *Bacteroides* isolates are clustered together under one umbrella called *Bacteroides fragilis* group that consists of 10 species, namely *B. distasonis*, *B. caccae*, *B. eggerthii*, *B. fragilis*, *B. merdae*, *B. vulgatus*, *B. ovatus*, *B. stercoris*, *B. thetaiotaomicron*, and *B. uniformis*. They all hydrolyze esculin and grow in the presence of 20% bile salt. Members of the *B. fragilis* group are resistant to colistin, kanamycin, and vancomycin and grow on media specially formulated for anaerobic bacteria.

CALYMMATOBACTERIUM GRANULOMATIS

Calymmatobacterium granulomatis, formerly known as *Donovania granulomatis*, are facultative anaerobic, Gram-negative bacilli that are believed to be closer to *Klebsiella* spp. because of their genomic similarities. They are known to cause granuloma inguinale, which is a chronic disease involving subcutaneous tissues in the general vicinity of the genital, anal, and inguinal areas. The infection is some-what rare in the United States, but not uncommon in tropical countries. The laboratory diagnosis can be made by demonstration of the bacterium in histiocytes in Giemsa or Wright stained smears. Cultures are usually not needed but can be made on coagulated egg yolk medium. Granuloma inguinale can be treated with tetracycline or macrolides.

CARDIOBACTERIUM HOMINIS

Cardiobacterium hominis strains are slow growing, Gram-negative bacilli that are facultative anaerobes. They are normally present in the upper respiratory tract from where they can enter the blood and attach to heart tissue. Though they usually appear in a rather low number, *C. hominis* are known to cause endocarditis, mostly in patients with preexisting conditions. Diagnosis is often based on its isolation in blood cultures at 35°C in the presence of 3%–5% carbon dioxide. Visible growth can be seen in 48–72 hours. Strains of *C. hominis* are generally sensitive to penicillin or ampicillin.

STREPTOBACILLUS MONILIFORMIS

Streptobacillus moniliformis are Gram-negative bacilli that are nonmotile, asporogenous facultative anaerobes. They are normally present in the nasopharynx of rats and known to cause rat-bite fever. The clinical condition may be accompanied by granules, bulbous swelling, and recurrent fever. The bacterium can be isolated from blood, joint fluid, and pus when cultured on media containing 15% defibrinated rabbit blood and incubated at 35°C in a carbon dioxide incubator. Most strains of *S. moniliformis* are sensitive to penicillins and tetracycline.

SPIRILLUM MINUS

Like *S. moniliformis*, *S. minus* is also a causal agent of rat-bite fever or spirillar fever, which follows bites by rats, mice, or some other animals including dogs and cats. The incubation period is usually 4 weeks, as opposed to 2 weeks for *S. moniliformis* infection. The clinical condition is characterized by inflammation, and induration, accompanied by lymphangitis. *Spirillum minus* cannot be cultured on most laboratory media. The diagnosis is generally based on the microscopic demonstration of the bacterium in the relevant clinical specimens, including the blood, exudates from the bite wound, or cutaneous eruptions near the site of the initial bite. These are Gram-negative small spirals that can also be stained with Giemsa stain. Direct examination of the clinical specimens using dark field microscopy is also useful. Also, like *S. moniliformis*, *S. minus* infections are treated with penicillin and tetracycline.

Chapter 11

Spirochetes and Bacteria without a Cell Wall

Spirochetes and bacteria without a cell wall do not quite fit in with the classic concepts of bacteria that have been discussed so far. It is also a fact that there is no similarity between the members of the two groups; they are very different from each other. They are discussed here in one chapter only for the sake of brevity.

SPIROCHETES

Spirochetes are spiral, Gram-negative bacteria with a unique mode of motility that is quite different from those of other bacteria (they lack external flagella). All bacteria classified as spirochetes generally have a helical protoplasmic cylinder made of a thin layer of peptidoglycan and a multilayered outer membrane. Spirochetes differ considerably from each other with respect to habitats and physiological characteristics. Three genera are associated with serious diseases in humans. These are *Treponema*, *Borrelia*, and *Leptospira*.

Treponema pallidum

Genus *Treponema* is known to have more than 10 species; most of them are anaerobes and are generally present in the mouth. Only two species, *T. pallidum* and *T. carateum*, are known to cause diseases in humans. Of the two, T. *pallidum* is perhaps the most important human pathogen. *Treponema pallidum* has been further divided into three subspecies, *T. pallidum* subsp. *pallidum*, the causal agent of syphilis; *T. pallidum* subsp. *pertenue*, which causes yaws; and *T. pallidum* subsp. *endemium*, which is mostly associated with nonvenereal syphilis. Strains of *T. pallidum* are ultra-small bacteria, about 0.1 μm in diameter but several microns in length. They are obligate parasites and they cannot live outside the host tissue. The primary focus of this chapter is on the etiologic agent of syphilis in humans.

A Concise Manual of Pathogenic Microbiology, First Edition. Saroj K. Mishra and Dipti Agrawal.
© 2013 Wiley-Blackwell. Published 2013 by John Wiley & Sons, Inc.

Disease

Treponema pallidum is the primary causal agent of syphilis, which is an important sexually transmitted disease and presents a global challenge. Approximately 30,000–40,000 persons are diagnosed with syphilis each year in the United States alone.

The three stages of syphilis

Primary   Characterized by formation of chancres (a localized lesion, about 1 cm in diameter) that usually develop at or near the site of the primary contact within 7–10 days. The lesion may heal spontaneously within 2–3 weeks.

Secondary   Secondary syphilis is characterized by diffuse rashes, including ones on the palms and soles, condyloma lata, and invasion of the central nervous system. Hepatitis and lymphadenopathy may also develop.

Tertiary   Tertiary stage develops after 1–3 years, in about 10%–20% of the untreated cases. Symptoms involving the central nervous system (CNS), cardiovascular system, and skin (gummas) may ensue, causing significant morbidity and mortality.

Lab Diagnosis

The causal agent, *T. pallidum*, cannot be cultured. Serological tests including ELISA and direct microscopic examination using dark field microscopy or fluorescence microscopy can be helpful. The bacterium appears as a thin spiral structure against a dark background (Fig. 11.1).

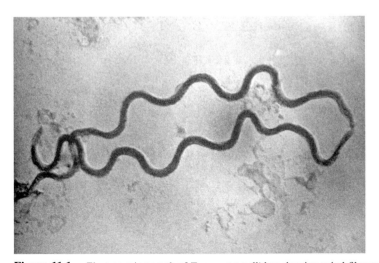

Figure 11.1.   Electron micrograph of *Treponema pallidum* showing spiral filaments (source: CDC).

Antibiotic Sensitivity

Most strains of *T. pallidum* are quite sensitive to penicillin. From the 1940s through the 1960s, the classic penicillin remained quite effective in the treatment of syphilis. However, penicillin-resistant strains have lately emerged, necessitating alternate therapeutic measures. Ceftriaxone and doxycycline can be used in primary and secondary disease cases, but penicillin still remains the drug of choice for the management of tertiary disease.

Leptospira interrogans

The genus *Leptospira* is comprised of two species, *L. interrogans* and *L. bifexa*. Only members of *L. interrogans* are human pathogens. On the basis of their antigenicity, *L. interrogans* strains can be further divided into several serovars. However, less than a dozen serovars are associated with human diseases.

Disease

Leptospira interrogans cause leptospirosis, primarily a zoonotic disease. Several wild animals including mongooses, jackals, foxes, raccoons, bats, lizards, and rats are believed to be carriers. Domestic animals including dogs, goats, cattle, pigs, sheep, and horses are believed to contract leptospirosis from wild animals. Leptospirosis is often responsible for serious morbidity and mortality in domesticated animals. The spirochete is usually present in their urine. Humans acquire leptospirosis through direct contact with infected animals or via inhalation of their aerosolized urine or exposure to water contaminated with their fecal matters. *Leptospira interrogans* can survive in soil for up to 2 weeks and in water for several months. The usual portal of entry is the mucous membrane of the mouth, eyes, and genitals.

Leptospirosis can be mild and self-healing or acute and fulminating. The incubation period is generally 5–10 days and the organ systems most commonly involved are central nervous system, liver, and kidneys. Symptoms during the fulminating stage include high fever, occasionally rising to 105°F (41°C), severe headache, myalgia, and malaise, which may be accompanied by nausea and vomiting. Serious liver and kidney damage may ensue. The disease can occasionally be fatal if untreated.

Laboratory Diagnosis

The laboratory diagnosis of leptospirosis requires a careful consideration of the clinical phase of the infection. During early stages, most relevant clinical specimens are blood and cerebrospinal fluid, but urine must be examined if the samples are to be collected during the later phase. Tight characteristic spirals are easily seen under direct microscopic examination using dark field microscopy (Fig. 11.2). *Leptospira interrogans* is easily cultured on media fortified with rabbit serum or bovine serum albumin with Tween 80. Such media are commercially available in the United States

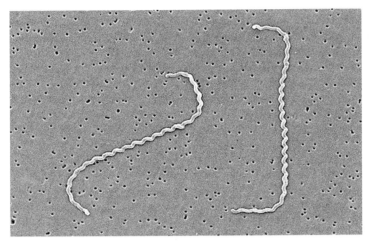

Figure 11.2.   Electron micrograph of *Leptospira interrogans*. Note the spirals are tighter than those seen with *T. pallidum* (source: CDC).

or can be made in the laboratory, if required. The incubation is done aerobically at 30°C for 10–15 days. Colonies of *L. interrogans* develop slowly and occasionally longer incubation period is required. Several serological tests including agglutination test are also available and quite useful.

Antibiotic Sensitivity

Antibiotic therapy is not always indicated as the disase is often self-limited. Most of the pathological conditions are apparently due to a strong immune response. If required, tetracycline, penicillin, and cephalosporins are useful antibiotics for the management of leptospirosis.

Borrelia Species

The genus *Borrelia* has several species, of which nearly 15 are indicated in human diseases, *B. burgdorferi* being the best known and perhaps the most important. *Borrelia* infections are transmitted by arthropods; therefore it is classified as a zoonotic disease. *Borrelia burgdorferi* is the causal agent of Lyme disease, which is a tick-borne infection. Its natural habitat is the *Ixodes* tick that infests field mice and deer. Another species, *B. recurrentis*, causes relapsing fever and is transmitted to humans by a louse species, *Pediculus humanus*. Other Borrelia infections are also tick-borne and are mostly confined to specific geographic regions. For example, *B. crocidurae* and *B. persica* are considered causal agents of tick-borne relapsing fever in North Africa and Asia, respectively.

Disease

Cases of Lyme disease caused by *B. burgdorferi* have been reported from the Americas, Europe, Asia, and parts of Africa. One week following the tick bite, erythema (an elevated lesion), extending up to several centimeters in diameter, develops at the site of the bite. The infection spreads hematogenously, causing fever, headache, and neck pain. It can be fatal if not treated. About 10,000 cases are reported each year in the United States. The louse-born relapsing fever develops 2–15 days following the infection and is characterized by a sudden onset of fever, headache, and myalgia lasting for 5–10 days. The infection can be occasionally complicated by myocarditis. *Borrelia burgdorferi* and *B. recurrentis* produce toxins. *Borrelia recurrentis* strains are also known for antigenic shifts that are probably responsible for periodic febrile episodes.

Laboratory Diagnosis

Lyme disease is usually diagnosed on the basis of clinical symptoms and case history. Serological tests, though available, should be used only in conjunction with clinical and epidemiological data. Most strains grow well on Kelly's modified medium, also known as Barbour-Stoenner-Kelly medium, under microaerophilic conditions. Inoculated media are generally incubated at 35°C for 7–14 days, occasionally longer.

Antibiotic Sensitivity

Tetracycline derivatives are usually preferred in the clinical management of Lyme disease but cephalosporins are indicated if the central nervous system is involved.

BACTERIA WITHOUT A CELL WALL

These are somewhat unusual bacteria in the sense that they lack a cell wall. Members of two genera, *Mycoplasma* and *Ureaplasma*, are important human pathogens.

Mycoplasma Species

Mycoplasma spp. are found in a wide range of niches including plants, arthropods, animals, and humans. Since they lack a cell wall, they do not have a regular shape, but most are approximately 0.1 μm in diameter. Instead, they maintain the stability of their cytoplasmic membrane with the help of sterols, which are obtained in the form of cholesterol from their animal hosts. *Mycoplasma* spp. have a small genome, consisting of 0.5–1.5 megabases; therefore they are often called "minimal cells." There may be some dispute as to the exact number of species within the genus *Mycoplasma*, but four species are believed to be strongly associated with infections in humans. These are *M. fermentans, M. genitalium, M. hominis*, and *M. pneumoniae*.

Disease

Mycoplasma fermentans has been noted in the cases of pneumonia in otherwise healthy individuals. It is also believed to be a causal agent of nephropathy and certain forms of disseminated infection in patients primarily suffering from acquired immunodeficiency syndrome (AIDS). Strains identified as *M. genitalium* have been isolated from the human oropharynx and urethra and indicated in urethritis, pelvic inflammatory disease (PID), and even pneumonia-like symptoms. *Mycoplasma hominis* is often isolated from the urethra, cervix, and vagina and is believed to cause PID, pyelonephritis, and bacteremia. Instances of pneumonia and meningitis have also been noted in some cases. Of the four species, *M. pneumoniae* is perhaps the best known and most important pathogen. It is the major cause of pneumonia in children. The symptoms generally include mild fever, cough, and headache. It may account for almost 20% of all cases of pneumonia in children and the elderly. In addition, it is known to cause extrapulmonary infections including otitis and pharyngitis.

Virulence Factors

An important pathogenic trait of *Mycoplasma* spp. is their ability to adhere to the epithelial cells with the help of adherence proteins. One of the adherence proteins identified in the case of *M. pneumoniae* is a 168kD protein, which is found at the tip of the cell. In addition, hydrogen peroxide and superoxide, byproducts of mycoplasmal metabolism, also play a role in its pathogenesis.

Laboratory Diagnosis

Mycoplasma pneumoniae is a very slow-growing microorganism. Throat swabs or sputum are streaked on *Mycoplasma* agar. It must be noted that there is no single medium that is good for all clinically significant species of the genus *Mycoplasma*. Cultures take a long time to grow. ELISA and PCR are faster and more effective tools of laboratory diagnosis. Since a number of species belonging to the genus *Mycoplasma* can be isolated from clinical specimens, it may sometimes be necessary to differentiate them. Utilization pattern of glucose, arginine, and urea can be helpful. None of the clinically significant *Mycoplasma* spp. utilize urea (an important test for differentiating them from *Ureaplasma*). *Mycoplasma pneumoniae* and *M. genitalium* utilize glucose but not arginine. On the other hand, *M. hominis* utilizes arginine but not glucose while *M. fermentans* utilizes both. Another species, *M. spermatophilum*, can be isolated from urethra, but seems to have an uncertain etiologic role, and utilizes neither glucose nor arginine.

Antibiotic Sensitivity

Since mycoplasmas are devoid of a cell wall, antibiotics that target the bacterial cell wall are not indicated. Erythromycin and tetracycline derivatives are quite effective against *M. pneumoniae* and possibly other species.

Ureaplasma urealyticum

Like *Mycoplasma* spp., *Ureaplasma* spp. are devoid of a cell wall. *Ureaplasma urealyticum* can be isolated from the cervix and vagina of nearly 75% of sexually active females, especially those from lower socioeconomic strata and those with multiple sex partners. It is seldom isolated from the genitals of prepubescent girls who have had no sexual contact with males. *Ureaplasma* spp. have a genome of 0.75–1.2 Mbp and a GC content of 27–30%. Six species have been recognized within the genus, but only *U. urealyticum* is of major clinical significance.

Disease

Even though *U. urealyticum* is frequently isolated from female genitals, it is not known to cause any infection in women. Low vaginal pH is suspected to play a role in this unusual phenomenon. In contrast, *U. urealyticum* is believed to be the causal agent in nearly 40% cases of nongonococcal urethritis in males. *Ureaplasma urealyticum* is also an important causal agent of pneumonia and respiratory tract infections in newborn, mostly in infants with low birth weight.

Laboratory Diagnosis

Ureaplasma is a fastidious and slow-growing bacterium. It requires urea for its growth. Ureaplasma agar is a selective medium and culture is perhaps the most dependable tool for the diagnosis of *Ureaplasma* infections.

Antibiotic Sensitivity

Tetracyclines, macrolides, and fluoroquinolones are effective agents against *U. urealyticum*.

Chapter 12

Actinomycetes

Actinomycetes are a group of metabolically dynamic microorganisms. They are better known for the production of a wide range of antibiotics, anticancer drugs, and industrial enzymes. Only a small number of species belonging to the order Actinomycetales are pathogenic. From the evolutionary perspective, actinomycetes form a bridge between the classic bacteria and fungi. The properties that mandate their classification with bacteria include prokaryotic cell structure and *in vitro* sensitivity to most antibacterial antibiotics. On the other hand, their tendency to form true branching with septate mycelium, external spores, and granulomatous tissue reaction in infected hosts bring them closer to fungi. The term "actinomycete" literally means "ray fungus." Because of their fungus-like properties, actinomycetes have been traditionally studied by mycologists. By early 1970s, it became obvious that these are prokaryotic microorganisms, and most mycologists started avoiding them, but bacteriologists did not quite embrace them either. As a result the study of actinomycetes has fallen into a "no man's land."

Actinomycetes are broadly divided into two major groups, the anaerobic actinomycetes and the aerobic actinomycetes. The anaerobic actinomycetes have only two medically important genera, *Actinomyces* and *Rothia*. The aerobic actinomycetes represent a much bigger group and include five genera of clinical significance that include species of *Nocardia, Actinomadura, Streptomyces, Thermoactinomyces,* and *Saccharopolyspora.* Aerobic actinomycetes that are commonly considered nonpathogenic include members of the genus *Acinoplanes, Streptomyces* (most species), *Streptosporangium, Micromonospora, Micropolyspora,* and several species of *Nocardia* and *Actinomadura.* Incidentally, the "nonpathogenic" aerobic actinomycetes are of immense industrial importance (see publications by Mishra et al. in the bibliography).

A Concise Manual of Pathogenic Microbiology, First Edition. Saroj K. Mishra and Dipti Agrawal.
© 2013 Wiley-Blackwell. Published 2013 by John Wiley & Sons, Inc.

ANAEROBIC ACTINOMYCETES

Actinomyces Species

The genus *Actinomyces* includes anaerobic species that are typically Gram-positive. Infections caused by *Actinomyces* spp. are commonly referred to as actinomycosis, a disease that was quite common during the pre-antibiotics era. It seems, due to widespread use of antibiotics, that actinomycosis has been almost inadvertently eradicated from much of the world. The genus *Actinomyces* has two pathogenic species. *Actinomyces israelii* is a human pathogen that causes systemic or localized infection. The systemic infection may involve the lungs or even the brain. Localized infections are chronic in nature and frequently involve the gums or cheeks. The other species, *A. bovis*, causes lumpy jaw in the cattle.

Laboratory Diagnosis

Pus or sputum are inoculated into thioglycollate broth and incubated anaerobically for about 1 week at 35°C.

Antibiotic Sensitivity

Penicillin and a number of other commonly used antibiotics are quite effective in the treatment of actinomycosis.

Rothia dentocariosa

Rothia dentocariosa are commonly present in the human mouth, often below the gum line. They are anaerobic to facultative anaerobes, and are believed to play a role in the etiology of dental caries. Systemic, cutaneous, or subcutaneous infections are not known.

AEROBIC ACTINOMYCETES

Nocardia Species

Unlike *Actinomyces* spp., *Nocardia* are aerobic bacteria and are commonly present in soil. Species of the genus *Nocardia*, especially *N. asteroids* and *N. farcinica*, and occasionally *N. otitidiscaviarum (N. caviae)* and N. *brasiliensis*, cause nocardiosis in humans. Nocardiosis is an airborne disease that primarily involves the lungs from where it may spread to other organs, especially to the brain for which it seems to have a predilection. Symptoms mimic tuberculosis and in fact in many parts of the world, nocardiosis is often referred to as Pseudotuberculosis. Symptoms may include cough, often accompanied by blood tinged sputum, and fever, which tends to be higher compared with tuberculosis. Like actinomycosis, cases of nocardiosis have greatly declined as compared with the pre-antibiotics era. However, in underdeveloped countries, it is quite possible that a significant number of tuberculosis patients actually suffer from nocardiosis.

Nocardia caviae and *N. brasiliensis* are better known for causing a clinically distinct condition called actinomycetoma, occasionally also referred to as "Madura foot." It must be noted here that the term mycetoma refers to a localized chronic swelling that is also caused by fungi (therefore, the term "actinomycetoma" is used here). The disease mostly involves the foot and lower leg. In the later stage, the swellings start draining with pus containing "sulfur granule," which is essentially a tight cluster of filaments surrounded by a large number of reactive cells, mostly neutrophils. Two other species that were previously classified as *N. madurae* and *N. pelletieri*, but are now called *Actinomadura madurae* and *Actinomadura pelletieri*, also cause actinomycetoma. No specific virulence factors have been identified except for the fact that a large number of aerobic actinomycetes produce toxins or antibiotic-like substances. Some of the key physiological differences between clinically significant aerobic Actinomycetes are noted in Table 12.1.

Laboratory Diagnosis

Laboratory diagnosis of nocardiosis is not easy. Because the causal agents are sensitive to antibiotics, antibiotics cannot be incorporated in the isolation media. But the clinical specimens are often contaminated with fast-growing bacteria against which the slow-growing *Nocardia* cannot compete. In 1969, Mishra and Randhawa developed a selective technique called Paraffin Bait Technique for the isolation of pathogenic *Nocardia* spp. from heavily contaminated specimens. It depends on the ability of nocardiae to use paraffin as energy source (Fig. 12.1). Once isolated, *Nocardia* can be cultured on modified Sabouraud agar (2% glucose, 1% neopeptone, 2% agar, 1 L water). Relatively uncontaminated clinical specimens, such as biopsied tissue and CSF, can be directly cultured on modified Sabouraud agar. Strains of *N. asteroides* show remarkable differences in their appearance (Fig. 12.2), and readers are referred to the classic works of Mishra and Gordon (1980, 1981). Also, since several microorganisms including members of the genus *Rhodococcus* show a remarkable morphological similarity with pathogenic *Nocardia* spp., a number of physiological tests must be conducted for specific identification (see Table 12.1).

Antibiotic Sensitivity

Even though *Nocardia* spp. are quite sensitive to antibiotics *in vitro*, most strains do not respond favorably to most antibacterial antibiotics *in vivo*. The drug of choice for the treatment of nocardiosis is trimethoprim-sulfamethoxazole or carbapenems often supplemented with tetracycline.

Table 12.1 Some Important Physiological Differences among Clinically Significant Aerobic Actinomycetes (Modified after Mishra et al., 1980)

Physiological tests	Nocardia asteroides	Nocardia caviae	Nocardia brasiliensis	Actinomadura madurae	Actinomadura pelletieri	Streptomyces somaliensis
Casein*	Negative	Negative	Positive	Positive	Positive	Positive
Hypoxanthine*	Negative	Negative	Positive	Positive	Positive	Negative
Xanthine*	Negative	Negative	Positive	Negative	Negative	Negative
Tyrosine*	Negative	Negative	Positive	Positive	Positive	Positive
Cellobiose**	Negative	Negative	Negative	Positive	Positive	Negative
Erythritol**	Negative	Negative	Negative	Negative	Negative	Negative
Glycerol**	Positive	Positive	Positive	Positive	Negative	Negative

* Hydrolysis/decomposition.
** Acid production.

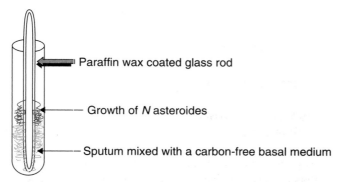

Paraffin wax coated glass rod

Growth of *N* asteroides

Sputum mixed with a carbon-free basal medium

Figure 12.1.   Schematic demonstrating the use of the paraffin bait technique for the isolation of pathogenic *Nocardia* from sputum. See color insert.

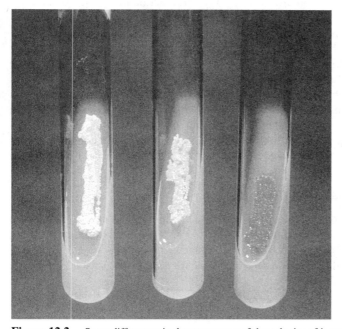

Figure 12.2.   Some differences in the appearance of the colonies of important aerobic actinomycetes: *N. asteroides* (left), *N. brasiliensis* (center), and *Rhodococcus* spp. (right). See color insert.

Streptomyces Species

Streptomyces somaliensis is the only unequivocally recognized species that is known to cause infection in humans. Isolation of strains later identified as *S. griseus* and *S. albus* from clinical specimens has been occasionally reported, but their actual role in the causation of any disease is doubtful. *Streptomyces somaliensis* has been recognized as the causal agent of mycetoma or Madura foot in several cases. The clinical signs and symptoms resemble those of mycetoma caused by other aerobic actinomycetes. Most strains grow well on Sabouraud agar.

THERMOPHILIC ACTINOMYCETES

Thermoactinomyces Species and *Saccharopolyspora* Species

Thermophilic actinomycetes grow at a higher temperature, usually >45°C, and they are commonly present in the core of straw bales and piles of hay, where the temperature can be in excess of 45°C. The disease variously called farmer's lung, allergic pneumonitis, or allergic alveolitis can be a life-threatening condition. It is essentially an allergic reaction, mediated by Type III reaction involving IgG and IgE antibodies. Cellular immunity also plays a role in the pathogenesis of this syndrome. Symptoms include pulmonary edema and breathing difficulties. The condition develops following repeated inhalation of the spores. Farmer's lung accounts for >12% of cases of hypersensitivity pneumonitis in the United States with an incidence rate ranging from 8 to 500 cases per 100,000 persons per year. In Europe, especially in the United Kingdom, the incidence rate is reportedly 400–2,500 per 100,000 persons per year in the farming community. The disease is also noted in other European countries including France, Sweden, and Finland. The thermoactinomycetes commonly incriminated in Farmer's lung include *Thermoactinomyces vulgaris, T. sacchari,* and *Saccharopolyspora rectivirgula.*

Laboratory Diagnosis and Control

The most dependable diagnostic test is the demonstration of specific precipitins in the immunodiffusion test. The patient's clinical condition often correlates to how strong the precipitin bands are. Effective therapeutic measures include avoiding exposure to thermophilic actinomycetes and a judicious use of cortisone.

Chapter 13

Introduction to Pathogenic Fungi and Superficial Mycoses

Fungi (singular fungus) are eukaryotes. They generally occur in two forms: yeast, which can be round or oval and basically unicellular, but capable of forming long chains called pseudomycelium; and the mold or filamentous. Some pathogenic fungi can be yeast-like inside animal tissue and filamentous in their natural habitat. Also, fungi have a highly developed form of sexual reproduction, but most can also multiply asexually. The natural habitat of a majority of fungi is soil where they perform their primary function in nature, that is, decomposing plant material and recycling the biomass in the ecosystem. However, certain pathogenic fungi are more frequently associated with pigeon or bat excreta. A majority of fungi are harmless to humans and animals. Only a small number of species are known to cause diseases in humans and animals though a majority of plant diseases are caused by fungi. The tissue reaction is usually granulomatous. Fungal infections do not respond to antibacterial antibiotics.

As stated in the introduction, taxonomically, fungi are divided into four divisions, namely Phycomycota (Phycomycetes) or Zygomycota (Zygomycetes), Ascomycota (Ascomycetes), Basidiomycota (Basidiomycetes), and Fungi Imperfecti (Deuteromycetes). But from the perspective of medical mycology, pathogenic fungi are divided into following three groups:

- Yeast-like fungi
- Dimorphic fungi
- Filamentous fungi

YEAST-LIKE FUNGI

Yeasts or yeast-like fungi can be round or oval, usually unicellular. Some species, such as *Candida* spp., form a multicellular chain of yeast-like cells that is called pseudomycelium (Fig. 13.1). Others, such as *Cryptococcus* spp., are usually

A Concise Manual of Pathogenic Microbiology, First Edition. Saroj K. Mishra and Dipti Agrawal.
© 2013 Wiley-Blackwell. Published 2013 by John Wiley & Sons, Inc.

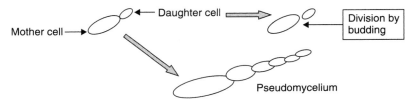

Figure 13.1. Depiction of budding and pseudomycelium formation by yeast-like fungi.

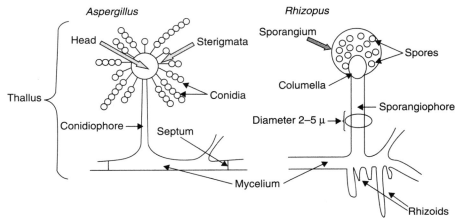

Figure 13.2. Illustration of the structure of two common molds.

spherical with a capsule. Most yeasts reproduce by asexual budding, but yeasts belonging to division Ascomycota demonstrate sexual reproduction. Yeasts typically multiply by a process called budding.

MOLDS OR FILAMENTOUS FUNGI

Filamentous fungi, also referred to as molds, are typically composed of a thallus, which includes mycelium, sporangiophore or conidiophore, spores, and a spore-bearing structure sometimes called the head. The mycelium may be septate as in the case of *Aspergillus* or aseptate as are *Mucor* and *Rhizopus* spp. (Fig. 13.2).

DIMORPHIC FUNGI

A unique group of fungi that exhibit properties of both the yeasts and the molds. A majority of dimorphic fungi are pathogenic. In their natural habitat, these fungi occur in their filamentous or mold form, but in host tissue or at human body temperature (37°C), they exhibit yeast form.

Diseases caused by fungi are broadly divided into following categories:

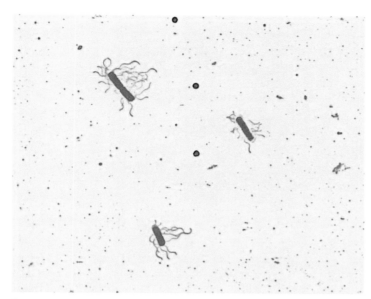

Figure 6.3. A smear from the culture of a *B. cereus* strain showing peritrichous flagella (Leifson flagella stain).

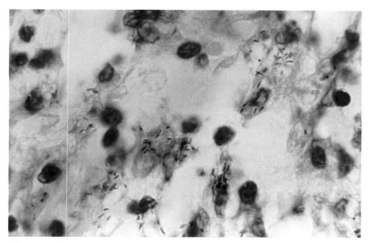

Figure 7.4. Acid-fast bacilli in a skin biopsy from a leprosy patient (source: CDC).

A Concise Manual of Pathogenic Microbiology, First Edition. Saroj K. Mishra and Dipti Agrawal.
© 2013 Wiley-Blackwell. Published 2013 by John Wiley & Sons, Inc.

Figure 8.2. A demonstration of the use of feeder bacterium *S. aureus* for the isolation of *H. influenzae*.

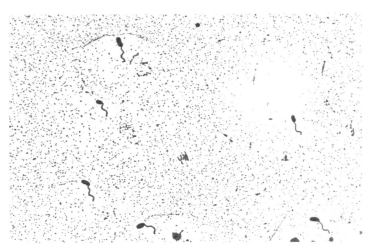

Figure 10.2. Photomicrograph of *Vibrio cholerae* showing a single polar flagellum in a digitally colorized Leifson flagella stain (source: CDC).

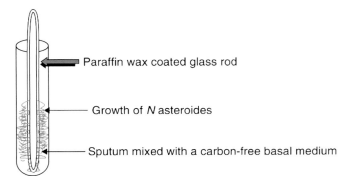

Paraffin wax coated glass rod

Growth of *N* asteroides

Sputum mixed with a carbon-free basal medium

Figure 12.1. Schematic demonstrating the use of the paraffin bait technique for the isolation of pathogenic *Nocardia* from sputum.

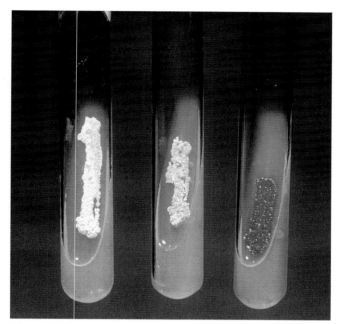

Figure 12.2. Some differences in the appearance of the colonies of important aerobic actinomycetes: *N. asteroides* (left), *N. brasiliensis* (center), and *Rhodococcus* spp. (right).

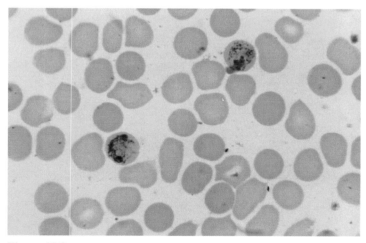

Figure 15.3. A blood smear stained with Giemsa stain depicting two red blood cells with *Plasmodium malariae* schizonts.

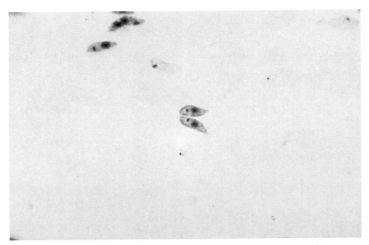

Figure 15.4. *Leishmania donovani*, leptomonad forms (courtesy: CDC).

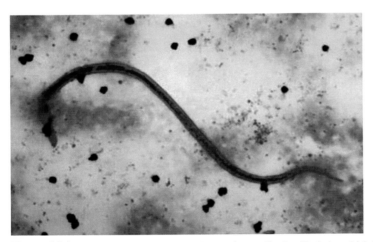

Figure 16.4. Photomicrograph of a *Wuchereria bancrofti* microfilaria in a thick blood smear using Giemsa stain (source: CDC).

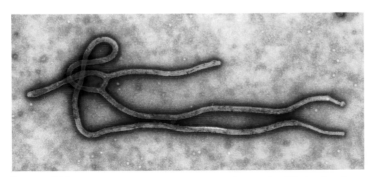

Figure 17.3. Colorized transmission electron micrograph of Ebola virus (courtesy: CDC).

Superficial Mycoses

The fungus infects mostly keratinized tissues, such as the epidermis, hair, and nails. The tissue invasion is minimal. We have also included infections involving the mucous membrane in this group. Certain forms of mucous membrane infections may involve some degree of tissue invasion. Most of the infections are acquired through close contact.

Subcutaneous Mycoses

Subcutaneous mycosis involves deeper layers of the skin and may eventually reach to the bones. The infections mostly result from the traumatic implantation of the pathogen into subcutaneous tissue. Most causal agents are soil-borne.

Systemic Mycoses

The etiologic agents of systemic mycoses are generally soil-borne and the respiratory tract is often (but not always) the portal of entry. Infections may involve any part of the body including the lungs, liver, kidneys, and brain.

For the sake of convenience, mycotic diseases have been divided into two chapters in this manual. This chapter includes a concise discussion on superficial mycotic disease, and the subcutaneous and systemic mycoses are summarized in the next chapter.

SUPERFICIAL MYCOSES

Black Piedra

The infection involves hair of the scalp. The fungus grows as a compact mass of cells forming a black nodule around the hair shaft. It may eventually invade the hair shaft. The size of the fungus mass may range from a few microns to greater than a millimeter in diameter. *Piedra hortae* is the main causal agent of black piedra. The infections are common in humid and warm countries. The laboratory diagnosis is mostly based on the microscopic examination of the hair shaft. Cultures can be made on Sabouraud agar on which it grows slowly and forms dark brown to black colonies. Multiple lines of treatment are available, but shaving off the hair appears to be the cheapest and most effective therapy.

White Piedra

White piedra is caused by *Trichosporon beigelii*, a yeast-like fungus that is considered dimorphic by some mycologists. The disease is characterized by white to creamy nodular fungal growth on the hair shaft. The fungus grows around and within the hair and mostly involves hairs of scalp. Unlike black piedra which is mostly restricted to scalp hair, white piedra may involve hairs of the beard and pelvic region. Cases of white piedra are seen all over the world, but the disease is more common in tropical regions. The laboratory diagnosis is often based on the microscopic examination of infected hair and demonstration of the mycelium (2–4 μm in diameter) and arthroconidia. *Trichosporon beigelii* grows well on Sabouraud agar. Heaped, cream-colored, yeast-like colonies can be seen after 4 or 5 days. Hyphae are septate and tend to break into small fragment called arthroconidia. A number of commercial over-the-counter products are available for the treatment. Topical preparations containing azole derivatives are useful. As stated in the case of black piedra, shaving off the hair and observing good personal hygiene is an effective means of controlling white piedra.

Tinea Versicolor

Tinea versicolor, also known as pityriasis, is caused by a yeast-like fungus, *Malassezia furfur*. The fungus is normally present on human skin. Infection mostly occurs on the upper torso and other areas rich in sebaceous glands. The infection is characterized by altered skin pigmentation that may look lighter in color than the normal skin. No tissue invasion or inflammatory reaction is observed. Tinea versicolor occurs worldwide, but more commonly in tropical countries. The fungus is lipophilic and requires olive oil or any other vegetable oil for growth. Diagnosis is mostly based on clinical examination but the pathogen can be cultured on Sabouraud agar supplemented with olive oil and incubated at 35°C. Direct microscopic examination of skin using a 10% sodium hydroxide solution can be helpful. Spherical to elongated yeast-like cells can be easily seen. The condition can be treated using an aqueous solution (20% w/v) of sodium hyposulfite, or topical preparations containing azole derivatives. However, the recovery to normal skin is a long, drawn out process.

Dermatophytoses

Dermatophytosis (plural: dermatophytoses), also known as tinea or ringworm, is a clinical condition involving keratinized tissues including skin, nail, and hair. The causal agents usually belong to genera *Trichophyton, Microsporum*, and *Epidermophyton*, and are collectively called dermatophytes. The term "dermatophyte," meaning skin-tree, goes back to the era when fungi were considered plants. Most dermatophytes produce keratinase, a proteolytic enzyme that hydrolyzes keratin. Therefore, many mycologists also refer to dermatophytes as keratinophilic fungi. With reference to their predilection and natural habitat, the dermatophytes are divided into three groups:

- **Anthropophilic:** These are primarily isolated from human sources and include species belonging to the genus *Trichophyton* (*T. rubrum, T. tonsurans,*

T. violaceum and others), *Microsporum (M. audouinii)*, and *Epidermophyton (E. floccosum)*.

- **Zoophilic:** Animals are the primary reservoir of zoophilic dermatophytes. Important members include species of *Microsporum (M. canis* and *M. nanum)* and *Trichophyton (T. mentagrophytes* and *T. verrucosum)*.
- **Geophilic:** Soil is the primary reservoir of geophilic dermatophytes. Notable species include *Microsporum gypseum* and *Trichophyton terrestre*.

Infections are acquired by direct contact with infected humans or animals or exposure of the bruised skin to soil. Dermatophytoses occur all over the world but they are more common in developing countries. There is a direct relationship between good personal hygiene and contracting dermatophytoses. A daily shower with generous soap application has dramatically reduced occurrences of dermatophytoses in much of the world. The lesions are initially localized, highly inflamed, and pruritic. The infection may remain localized, forming a round ring-like lesion, or may spread fast, covering a large area of skin. Some of the common clinical conditions along with the dermatologists' classifications of dermatophytoses are summarized below:

Tinea Corporis

Tinea corporis, also known as ringworm of the glabrous skin, is perhaps the most common form of dermatophytoses. The infection is limited to skin and the lesions are characterized by inflammation, erythema, and vesicle formation. Species most frequently involved include *T. rubrum, T. mentagrophytes*, and *M. canis*.

Tinea Capitis

Tinea capitis refers to scalp infection. The lesions are characterized by inflammation, ulceration, and hair loss. *Microsporum canis* and *T. tonsurans* are most frequently involved in this clinical type.

Tinea Barbae

Tinea barbae is characterized by pustular folliculitis mostly affecting the hairs of the beard. Species commonly involved include *T. rubrum, T. violaceum*, and *M. canis*.

Tinea Pedis

Tinea pedis, also known as athlete's foot, refers to infection involving toe-webs and sole. Infections are frequently associated with persistent moist conditions. The lesions are scaly, erythematous, and inflamed. Species most commonly involved include *T. rubrum* and *T. mentagrophytes*.

Tinea Unguium

Tinea unguium, also known as onychomycosis, refers to a clinical condition involving the nails. The infection may involve the nail-plate and spread to the area under the nail, resulting in a total deformation and loss of the infected nails. Species most commonly associated with tinea unguium include *T. rubrum* and *T. mentagrophytes*. Species of the yeast-like fungus *Candida* are also known to cause onychomycosis. Therefore, a differential diagnosis is required because of the differences in the effective therapeutic measures.

Laboratory Diagnosis

Dermatologists tend to diagnose dermatophytoses on the basis of affected sites and clinical symptoms. However, a definitive diagnosis requires isolation of the causal agents in culture. For a presumptive diagnosis, samples of skin scrapings are mixed with a 10% solution of sodium hydroxide and examined directly under a bright field microscope. The nail specimens should be cut into finer pieces and a 20% sodium hydroxide solution should be used. Fungal elements, if present, can be easily seen. Dermatophytes grow well on Sabouraud agar fortified with chloramphenicol (50 mg/L) and cycloheximide (500 mg/L). Chloramphenicol will suppress bacterial growth and cycloheximide is toxic to most of the environmental fungi that are commonly present on the skin and nails as transient microbiota. Inoculated plates should be incubated at 25°C. Visible colonies appear within 1 week but sporulation may require a longer period of incubation, often up to 2–3 weeks. Species are identified on the basis of colony appearance and size and shape of conidia. *Microsporum* and *Trichophyton* spp. tend to produce two types of conidia, the macroconidia and the microconidia. Size and shape of macroconidia is most useful in species identification. The macroconidia of *Microsporum* are fusiform or spindle-shaped (Fig. 13.3), thick-walled, and measure 7–20 × 35–120 µm, and those of *Trichophyton* are clavate with smooth walls and measure 4–8 × 8–50 µm. The macroconidia of *Epidermophyton* spp. are widely clavate with smooth walls, rounded distal ends, and measure 6–10 × 8–15 µm. The macroconidia of all the three species are multinucleate. Sexual stage (perfect stage) has been recognized in many species.

Antibiotic Sensitivity

Ever since it was introduced in 1950s, griseofulvin has remained a drug of choice for the treatment of dermatophytoses. Griseofulvin is given orally and the dose and duration of the therapy varies widely. However, cases of drug resistance to griseofulvin are on rise and most dermatologists prefer newer azoles. It must be noted that the griseofulvin is not effective against *Candida* infection. Multiple lines of topical preparations are also available and reportedly effective in some cases.

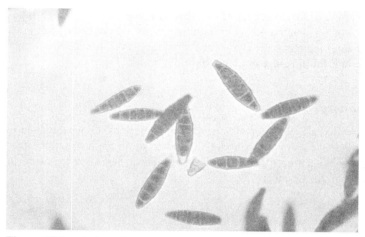

Figure 13.3. A photomicrograph showing spindle-shaped macroconidia of *Microsporum gypseum* in a laboratory-grown culture stained by Lactophenol cotton blue (source: CDC).

Otomycosis

Otomycosis, also known as mycotic otitis externa and fungal ear infection, is a superficial chronic or subacute infection of the outer ear canal. Otomycosis is a common infection, reported from all over the world. Symptoms frequently noted include inflammation, scaling, pruritus, and pain. Partial deafness may ensue in some cases due to occlusion of the ear canal by plug of fungal hyphae and epithelial debris. In many cases, the inflammation and scaling extends all the way through ear canal. In most cases, the tympanic membrane is not perforated. Unlike bacterial otitis, which is characterized by foul smelling pus discharges, otomycosis is nonexudative.

The causal agents are common airborne fungi. The species most frequently incriminated belong to the genus *Aspergillus* and include *A. niger, A. terreus* and *A. fumigatus.* Species of other fungal genera, such as *Penicillium, Scopulariopsis, Mucor, Rhizopus, Candida*, and dermatophytes are occasionally indicated as well.

Laboratory Diagnosis

Otomycosis can be differentiated from bacterial otitis by its characteristic nonexudative symptoms. Confirmatory test can be performed by microscopic examination of ear plugs and scales. Hyphae, sporulating vesicles, and fungal spores are easily demonstrated. If too much cellular debris or wax is present, the specimen can be digested in 10% sodium hydroxide solution. Cultures can be notoriously unreliable in the absence of clearly formed hyphal plugs. If plugs can be obtained, they should be cultured on Sabouraud dextrose agar fortified with chloramphenicol and incubated at 35°C. Once again, slanted media in large test tubes are preferred over Petri dishes, because the later increases the risk of contamination. A simultaneous culture for bacteria can be helpful in ruling out bacterial infection.

Antibiotic Sensitivity

Otomycosis is easily treated with topical antifungal preparations. Older therapeutic choices containing 0.1%–1% thymol or iodochlorhydroxyquin are also quite effective. Keeping the ear canal free of moisture buildup and drying after exposure to water are helpful prophylactic means.

Mycotic Keratitis

Mycotic keratitis is usually manifested as a corneal ulcer and with the presence of pus-like fluid in the anterior chamber of the eye (hypopyon). The corneal ulcers are mostly raised with a white to greyish appearance and irregular margins. Other features may include delicately radiating lines at the perimeter anterior stromal infiltrate and satellite lesions. If not treated in a timely manner, mycotic keratitis can lead to blindness. The disease appears to be more common in farmers and laborers exposed to dust and prone to what might be called minor eye injuries. The rising incidence rate of mycotic keratitis is generally attributed to uninhibited use of eye drops containing antibacterial antibiotics and corticosteroids. The risk of infection is also higher if the patient has had an abraded cornea, or the eye has been subjected to surgical procedures. Glaucoma, corneal diseases, and corneal trauma are generally believed to be important predisposing conditions. The most important causal agents of mycotic keratitis are *Fusarium solani, F. oxysporum*, and *F. nivale*. Other species of *Fusarium, Aspergillus fumigatus, Curvularia lunata, Penicillium citrinum, Phialophora verrucosa*, and *Candida albicans* have also been occasionally implicated. Most of these fungi are normally present in soil and some (such as *F. solani*) are important plant pathogens. Their isolation from soil is fairly common.

Laboratory Diagnosis

A dependable laboratory diagnosis would require repeated isolation of the fungus from carefully collected clinical specimens obtained from the affected areas. A direct microscopic examination of corneal scrapping can be quite helpful.

Antibiotic Sensitivity

Mycotic keratitis has been successfully treated with topical application of amphoteric B and Pimaricin solutions. Pimaricin is believed to be more effective in the treatment of infection caused by *Fusarium* spp.

Table 13.1   Some Delineating Characteristics of Yeast-like Fungi Commonly Associated with Mucocutaneous Infections

Species	Germ tube formation	Galactose*	Inositol*	Maltose*	Trehalose*	Xylose*
Candida albicans	Positive	Positive	Negative	Positive	Negative	Negative
Candida parapsilosis	Negative	Positive	Negative	Positive	Positive	Positive
Candida tropicalis	Negative	Positive	Negative	Positive	Positive	Positive
Torulopsis glabrata	Negative	Negative	Negative	Negative	Positive	Negative

* Utilization as sole source of carbon.

MUCOCUTANEOUS MYCOSES

Several pathogenic fungi, especially dimorphic fungi, are known to infect mucous membranes, often as a secondary complication. In this subsection, we will focus only on the mucous membrane infections by *Candida* spp. Some of the differentiating features of yeast-like fungi associated with mucocutaneous infections are depicted in Table 13.1. The common clinical conditions caused by *Candida* and related species can be divided into two distinct categories.

Thrush

Thrush is characterized by curd-like growth of the yeast on tongue and palate. A confluent growth of yeast-like cells and pseudomycelium can form a biomembrane on the surface. The causal agent is frequently *Candida albicans*, but other species are also occasionally incriminated. Thrush mostly occurs in newborns, who acquire the disease during passage through the birth canal. It can also be noted in immune-compromised patients, such as those with AIDS, persons on steroids, and diabetics. The condition is often indicative of host's immune status. Thrush is frequently diagnosed on the basis of clinical symptoms. A definitive diagnosis can be made by direct microscopic examination of swabs or tongue scrapings, which would reveal yeast-like cells and abundant pseudomycelium. Cultures can be made on Sabouraud agar containing chloramphenicol. Most *Candida* spp., except *C. albicans*, are sensitive to cycloheximide. Thrush can be also an annoying complication in stomatitis. Topical medications are often sufficient to treat thrush but systemic treatment may be needed for more severe cases, most commonly with azoles.

Vulvovaginitis or Vaginitis

Vulvovaginitis or vaginitis is commonly referred to as yeast infection. The disease is characterized by the inflammation of vagina, labia, and surrounding areas. The symptoms may include burning, itching, and curd-like odorless vaginal discharge with painful intercourse. In some cases, as a complication of vaginitis or as a disease by itself, *Candida* spp. can cause urinary tract infection (UTI), especially urethritis, which can be painful and chronic. Vulvovaginitis is a common complication during the late stage of pregnancy. The primary cause of infection appears to be loss of resident microbiota in vagina, often due to prolonged antibiotic therapy and occasionally due to hormonal changes. Vulvovaginitis is also a common infection in postmenopausal women. The disease is sexually transmissible. In males, the infection may result in balanitis characterized by inflammation of the glans of penis, accompanied by a burning and itching sensation. However, in most males the infection is mild and frequently unnoticed. Asymptomatic males can play a role in transmitting the disease to females. *Candida* vulvovaginitis occurs all over the world but the incidence rate seems to be higher in industrialized countries, perhaps due to excessive use of douche and other tools of intimate hygiene, which can disturb or even dislodge the resident microbiota. Though not a mucocutaneous infection, *Candida* spp. can also cause perianal infection in males and females. No well-defined virulence factor is known, but many strains of *Candida albicans* are known to produce proteinases and possibly certain extracellular toxins, which probably play important roles in the inflammation and tissue damage.

Laboratory Diagnosis

A vaginal swab can be cultured on Sabouraud agar supplemented with chloramphenicol. Inoculated plates should be incubated aerobically at 35°C. In some chronic cases of vaginitis, *Candida* spp. tend to form a biofilm which makes it hard to dislodge and isolate the fungus in culture. Species identification may not be considered essential for therapy, but it can be achieved by some simple tests and enforced by physiological tests including carbohydrate assimilation tests (see Table 13.1). *Candida albicans* forms a germ tube in serum and egg albumin, and produces chlamydospores on cornmeal agar (Fig. 13.4). *Candida albicans* is the principal causal agent of thrush and vulvovaginitis, responsible for infection in 70%–80% of cases. During the past few decades *Torulopsis glabrata* (formerly called *Candida glabrata*) and *Candida tropicalis* have emerged as important causal agents, especially in vaginitis and UTI.

Antibiotic Sensitivity

Therapy of *candida* vaginitis has a checkered history. In the pre-antibiotic era, the infection was mostly treated with Gentian violet (crystal violet) or by a long-term and generous application of natural yogurt on and in the vagina. Because Gentian violet leaves an intense violet/purple color that lasts for many days, its usage was

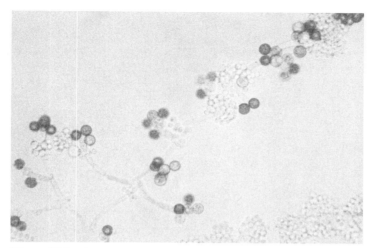

Figure 13.4.   Photomicrograph of a *C. albicans* culture on cornmeal agar containing Tween 80 and showing chlamydospores (Lactophenol-Cotton blue stain).

discontinued after nystatin and amphotericin cream and suppositories became available. Amphoteric B, otherwise a fine drug, was discontinued mostly because of its yellow color and the fecal appearance of the vaginal discharges following suppository application. Currently azoles, most commonly fluconazole given as a single dose, are popularly in use, as are topical therapies. However, strains resistant to azoles are emerging. In those cases, amphotericin B cream or suppository may still be the best option, if they can be obtained. Resistance to this agent is virtually unknown and the toxicity is limited because the drug is not absorbed. The infection tends to recur unless the normal vaginal microbiota is restored. For this purpose, some physicians recommend external application of *Lactobacillus*-containing products, and the old standby yogurt is not a bad option.

Chapter 14

Subcutaneous and Systemic Mycoses

SUBCUTANEOUS MYCOSES

Infections defined as subcutaneous mycoses affect deeper layers of the skin with intense tissue reaction, often resulting into a chronic condition. Localized swelling, a characteristic feature of subcutaneous mycoses, may be moderate to heavy and may spread to involve other organs. Based on the causal agent and the clinical condition, subcutaneous mycoses are divided into various subsets, namely chromoblastomycosis, phaeohyphomycosis, sporotrichosis, lobomycosis, and mycetoma. Involvement of subcutaneous tissues as a localized infection or as a consequence of the primary systemic infection may be noted in several cases of deep mycosis and will be discussed in the next part of this chapter. Typically, most subcutaneous infections result from a traumatic implantation of the pathogen into the skin.

Chromoblastomycosis

Chromoblastomycosis refers to a syndrome initially characterized by localized swelling, which develops slowly as discolored scaly papules that increase in size. The older lesions become raised, bear a shade of grey, red, or violet, and assume a wart-like appearance. Secondary lesions often develop in nearby areas. Any part of the body may be involved but most infections are seen on the hand, lower leg, chest, and back. Some of the important causal agents that are collectively called dematiaceous fungi include *Fonsecaea pedrosoi* (*Phialophora pedrosoi*), *Phialophora verrucosa* (Fig. 14.1), *Cladosporium carrionii*, and *Rhinocladiella aquaspersa*. These are soil-borne fungi and they are frequently isolated from decaying wood and other vegetable matter. Cases of chromoblastomycosis have been reported from all over the world, but the disease is more common in manual workers in less developed countries. Involvement of internal organs, such as the lungs and brain, is rarely seen.

A Concise Manual of Pathogenic Microbiology, First Edition. Saroj K. Mishra and Dipti Agrawal.
© 2013 Wiley-Blackwell. Published 2013 by John Wiley & Sons, Inc.

Figure 14.1. Photomicrograph of *Phialophora verrucosa* grown in a slide culture (lactophenol cotton blue stain; courtesy: Dr. Libero Ajello).

Mycetoma

Mycetoma or Madura foot is a localized infection that mostly involves feet and other parts of the body such as the arm, back, or any site subject to traumatic implantation of the pathogen. The infection mostly involves skin and subcutaneous tissues and may eventually spread to fascia and bone. The lesions are characterized by swelling with granuloma, abscess formation, and draining sinuses. Pus-like material draining from the sinuses contains granules measuring from 100 μm to up to 2 mm. Granules consist of a compact mass of hyphae and reactive tissue cells. The disease has been reported in several parts of the world, but mostly Sudan and other North African countries. In fact, the first cases were reported from Madura, a place in South India, hence the name Madura foot. The causal agents mostly include *Pseudallescheria boydii* (its asexual stage is called *Scedosporium apiospermum*), *Madurella mycetomi* (*M. mycetomatis*), and occasionally *Phialophora (Exophiala) jeanselmei*, *Aspergillus nidulans*, and several other fungi. These fungi are commonly present in soil.

Laboratory Diagnosis

Since signs and symptoms of chromoblastomycosis and mycetoma may resemble similar infections by a number of other fungi, bacteria, and parasites, a differential diagnosis is required. It is, however, not easy to isolate the fungus in culture from chromoblastomycosis cases unless the biopsied tissues are examined histopathologically, or cultured on Sabouraud agar supplemented with chloramphenicol. Histopathological examination of the tissue may reveal pigmented (dark) fungal cells in the granulomatous lesions. In cultures, the fungus develops as compact, dark-colored colonies. In the case of mycetoma, a microscopic examination of pus granules can provide useful clues. Cultures can be made from the biopsied tissue or granules. Species recognition is based on the size and shape of conidia and their arrangement on the spore-bearing structures variously called phialides or conidiophore. Mycelia are septate and darkly pigmented. Certain species that cause chromoblastomycosis form yeast-like growth, hence they are also called black yeast.

Antibiotic Sensitivity

A combination of surgery and a judicious use of amphotericin B or azole derivatives can be useful in the management of chromoblastomycosis and mycetoma. However, mycetomas often require radical surgery, even amputation of the affected limb.

SYSTEMIC MYCOSES

The terms "systemic mycoses" (singular mycosis) or "deep-seated mycoses" refer to fungal diseases involving internal organs, such as the lungs, brain, and kidneys. However, systemic mycoses may also develop into cutaneous, subcutaneous, or mucocutaneous mycoses, mostly as a secondary, but occasionally as the sole manifestation. The causal agents of systemic mycoses may be broadly divided into three groups: the dimorphic fungi, yeast-like fungi, and filamentous fungi or molds. Systemic mycoses are classic examples of airborne diseases. Respiratory tract is the usual portal of entry but infections due to traumatic implantation are also known. Lungs are often the primary site of infection from where the disease may spread to other parts of the body, including the CNS, bones, and skin. The three groups of the pathogens are discussed below.

DISEASES CAUSED BY DIMORPHIC FUNGI

Dimorphic fungi are filamentous (mold-like) in their natural habitat, but yeast-like in human tissue. In the laboratory, cultures grown at 25°C show the filamentous stage and the yeast form is seen at 35°C. The laboratory diagnosis is mostly based on the isolation of the pathogen from a suitable clinical specimen, such as sputum, bronchial aspirate, or biopsied tissue. The samples are inoculated on Sabouraud dextrose agar and brain heart infusion agars fortified with chloramphenicol (penicillin and streptomycin are preferred in some cases). Generally, two sets of Petri dishes or slanted media (in test tubes) are used. One set is incubated at 25°C and the other at 37°C (some mycologists prefer 35°C). Serological tests are useful in some cases. No specific virulence factor is known in this group of fungi. Amphotericin B is the drug of choice in most cases. Important differences among common dimorphic fungi are listed in Table 14.1. Some of the well-known mycotic diseases caused by dimorphic fungi include the following:

- Blastomycosis
- Coccidioidomycosis

Table 14.1   Some Important Differences among the Pathogenic Dimorphic Fungi

Species	Infective propagule	Size of infective propagule	Tissue form	*Size of tissue form (yeast phase)
Histoplasmosis capsulatum	Micro- and macroconidia	Microconidia: 2–3 μm Macroconidia: 8–14 μm	Spherical to oval yeasts	3 μm
Blastomyces dermatitidis	Conidia	2–10 μm	Large cells with a thick wall	8–15 μm; some 20–30 μm
Coccidioides immitis	Arthrospores	2–5 μm	Spherule or sporangia with spores	Spherule or sporangium: 20–50 μ; spores: 2–5 μ
Paracoccidioides brasiliensis	Probably arthrospores	Unknown	Large spherule with daughter cells	Spherule: 12–30 μm; daughter cells: 2–4 μm
Sporothrix schenckii	Conidia (elliptical)	2–3 μm × 3–6 μm	Cigar-shaped bodies	1–3 μm × 3–10 μm

- Histoplasmosis
- Paracoccidioidomycosis
- Sporotrichosis

Blastomycosis

Blastomycosis is caused by *Blastomyces dermatitidis*. The natural habitat of the fungus is not unequivocally established, but soil is believed to be its natural reservoir. Infections are mostly reported from the Ohio River and Mississippi–Missouri River valleys, southern Canada, and the mid-Atlantic and Great Lake regions. Blastomycosis has been occasionally reported from Europe, Asia, and Africa. The disease is airborne and the respiratory tract is believed to be the usual portal of entry. However, the possibility of traumatic implantation of the fungus in patients with only skin involvement cannot be ruled out. In most cases, the disease is asymptomatic, but in some it may present symptoms of mild respiratory tract infection accompanied by low-grade fever. The disease is mostly confined to the lungs but dissemination is noted in some cases. In fulminating cases, which are rare, the fever may become chronic and be accompanied by productive cough, malaise, and weight loss. A widening of the hilar shadow can be seen during X-ray examination in the early phase of infection. The disseminated form may involve oronasal mucosa, cutaneous and subcutaneous tissues, bones, testes, the prostate, and the CNS. Cutaneous blastomycosis is chronic and slow progressing. Like all other cases of systemic mycoses, blastomycosis should be differentiated from other chronic granulomatous diseases, such as histoplasmosis, tuberculosis, silicosis and relevant forms of malignancies. Cases of blastomycosis have somewhat declined in the recent years. The fulminating form of blastomycosis can be fatal if not treated.

Laboratory Diagnosis

Sputum or homogenized biopsied tissues are inoculated in duplicate sets on Sabouraud agar and brain heart infusion agar. One set is incubated at 25°C and the other at 35°C for 1–3 weeks. Slanted media (in test tubes) are preferable because they minimize dehydration of the medium over time. At room temperature, colonies develop slowly as white mold. At microscopic examination, smooth-walled, round to oval conidia borne laterally or terminally and measuring 2–10 μm in diameter can be seen (Fig. 14.2). Large, thick-walled spherical cells measuring 8–15 μm in diameter, representing the yeast phase of the fungus are seen at the microscopic examination of the growth at 35°C. The perfect state of the fungus is *Ajellomyces dermatitidis*, which is an ascomycete (Ascomycota). There are no skin tests available and value of serological tests is equivocal. Histopathological examination generally requires staining the slides with Gridley stain, Gomori Methenamine-silver stain, or Periodic acid-Schiff stain. Histopathological diagnosis depends on the demonstration of the characteristic yeast phase of the fungus with a sharply defined cell wall of sufficient thickness to give an appearance of double contoured wall.

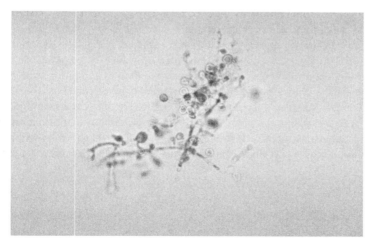

Figure 14.2.   Photomicrograph of the mycelial phase of *B. dermatitidis* showing typical conidia (lactophenol-cotton blue).

Antibiotic Sensitivity

As in most cases of systemic mycoses, amphotericin B, despite its high toxicity, is still the drug of choice for severe disease. Azole therapy with Itraconazole can be used for milder infection.

Coccidioidomycosis

The causal agent of coccidioidomycosis is *Coccidioides immitis*. Its natural habitat is alkaline soil in the semiarid regions, like those in southern California and Arizona where it grows in soil in its mold form. The mycelia frequently produce arthrospores (arthroconidia), which are easily airborne, especially during dust storms. The infection is airborne through the inhalation of arthrospores. The incubation period is 7–28 days, and the lungs are the primary site of infection. In most cases, the disease manifests symptoms of mild upper respiratory tract infections, which are often self-limited. Symptoms in acute cases include fever that can rise up to 40.5°C, productive cough with white, purulent, blood-tinged sputum, generalized aches, malaise, and weight loss. In its disseminated form, the CNS is frequently involved, with meningitis being the most common manifestation. Skin involvement, seen in some cases, includes erythematous lesions on thighs, knees, and other areas of body. Coccidioidomycosis is perhaps the commonest of all laboratory-acquired mycotic diseases.

Laboratory Diagnosis

Sputum, cerebrospinal fluid, and biopsied tissues should be cultured on Sabouraud agar and brain heart infusion agar. Test tubes rather than Petri dishes should be used. As in other cases of systemic mycoses, cultures should be made in duplicate and incubated at 25°C (room temperature) as well as 35°C. At room temperature, the colonies develop fast and sporulation can be seen in 7–10 days. The spores in mycelial phase called arthrospores or arthroconidia are barrel-shaped, borne laterally, and measure 2.5–4 × 3–6 μm. The yeast form of the fungus produces large spherules which measure 30–60 μm in diameter. At the histopathological examination of the biopsied tissues, spherules in various stages of development can be seen (Fig. 14.3). At maturity, the spherules develop a large number of internal spores called sporangiospores that measure 2–5 μm in diameter. Serological tests, such as agglutination tests and immunodiffusion tests, are useful diagnostic tools. Skin tests using mycelial phase antigen coccidioidin or yeast phase antigen spherulin are also helpful.

Antibiotic Sensitivity

Azole antibiotic fluconazole has been used successfully in most cases, but amphotericin B still remains the drug of choice for the management of coccidioidomycosis that does not respond to azole therapy, or for severe stages of the disease.

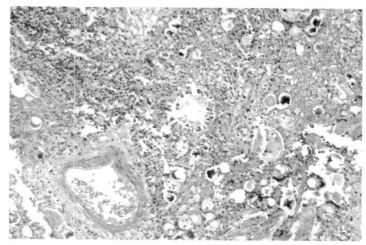

Figure 14.3.   Spherules of *C. immitis* in a biopsied tissue (methenamine-silver stain).

Histoplasmosis

Histoplasmosis is caused by *Histoplasma capsulatum*. The disease may be benign, subclinical, chronic, or fulminating. In most cases, the lungs are the primary site of infection from where it may spread to the liver, spleen, and bone marrow. Radiologically, the early stage is characterized by patchy shadows suggestive of pneumonitis which may develop into calcified fibrous tissue and the image may resemble that of miliary tuberculosis (bullet holes). The symptoms in the cases of pulmonary histoplasmosis usually include nonproductive cough, shortness of breath, mild-to-sever chest pain, and hemoptysis at a later stage. Other associated symptoms may include fever, night sweats, weight loss, and malaise. The yeast phase of the fungus mostly grows in the histiocytes, especially in the macrophages. Histoplasmosis is frequently seen as a secondary complication in the cases of AIDS. The causative agent, *H. capsulatum*, has been isolated from soil in many parts of the world. In the United States, it is known to occur in soil in the Mississippi and Ohio River valleys. The fungus seems to prefer soil enriched by chicken manure. Bats are considered natural carriers and the fungus is commonly found in bat excreta. Infections in immunecompetent individuals are often self-healing. Fulminating infections, if untreated, are mostly fatal. A variant called *H. capsulatum var. duboisii* is frequently associated with histoplasmosis in African countries.

Laboratory Diagnosis

Most strains of *H. capsulatum* grow well on Sabouraud agar and brain heart infusion agar. A visible growth may take 1–3 weeks. At 25°C, the growth is characterized by

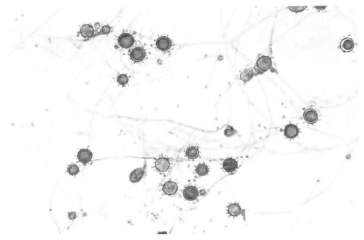

Figure 14.4.   Photomicrograph showing macroconidia of *H. capsulatum* in a culture grown at the room temperature.

whitish, fluffy colonies, representing the mycelial phase. Two types of conidia can be seen. The larger ones, called macroconidia, are spherical and measure 8–14 μm in diameter; the smaller ones, called microconidia, are also spherical and measure 2–4 μm in diameter (Fig 14.4). Histopathologically, spherical or oval yeast-like cells measuring 1–5 μm in diameter are mostly seen in the histiocytes. The tissue forms of *H. capsulatum var. duboisii* are much larger and measure 15 μm in diameter. The final confirmation requires phase conversion, which is the ability of the mycelia phase to grow into yeast forms at 37°C or vice versa. A skin test using histoplasmin can be useful in nonendemic areas. Immunodiffusion tests are also useful diagnostic tools.

Antibiotic Sensitivity

Amphotericin B, though toxic, is the most effective therapeutic agent available to date. No unequivocal instances of resistance to this antibiotic are reported. Itraconazole is often used in milder cases and for maintenance therapy in those with AIDS.

Paracoccidioidomycosis

Paracoccidioidomycosis, also known as South American blastomycosis, is caused by dimorphic fungus *Paracoccidioides brasiliensis*. The natural habitat of the fungus is probably soil, but none of the reports claiming its isolation from soil are without controversy. The disease is confined to South American countries, mostly to Brazil, and less commonly to other areas. The infection is believed to be airborne, but traumatic implantation of the fungus in bruised skin is also an important factor. Clinically, the disease manifests itself in several forms including mucocutaneous, cutaneous, lymphatic, visceral, and pulmonary infections. Oral lesions, probably the most common form of Paracoccidioidomycosis, involve oral mucosa, which may lead to the development of ulcerative stomatitis. Other forms of mucocutaneous legions mostly develop on nasal and anorectal mucosa. Skin infections mostly develop as crusted lesions on the face. The lymphatic lesions usually involve cervical lymph nodes which get enlarged and drain through the sinuses. The draining pus often contains a large number of fungal elements. The pulmonary infections are characterized by diffused infiltration, signs of obstruction, and occasionally cavity formation. The visceral form mostly includes infection of the liver, spleen, and intestines. Histopathological response includes granulomatous reaction with pyogenic inflammation and is generally similar to that seen in the cases of blastomycosis.

Laboratory Diagnosis

Pus from the draining sinuses, sputum, or biopsied tissue should be cultured on Sabouraud agar or brain heart infusion agar in duplicate and incubated at room temperature and at 35°C. The fungus grows slowly at 25°C and rarely produces any fructification. The yeast-like growth at 35°C is faster and more helpful in the mycological diagnosis. The yeast phase is characterized by large, spherical to oval spherule-like cells measuring 2–20 μm in diameter. In most instances, the entire surface of the spherule is surrounded by small buds measuring 2–4 μm in size. Histopathological examination of the biopsied tissue stained with hematoxylin-eosin (H&E) or any other fungus-specific stain generally reveals large spherical bodies (spherules) surrounded by external buds that may cover the entire surface. In the absence of the external buds, it may be hard to differentiate this fungus from the tissue phase of coccidioidomycosis, especially if the spherules are young and devoid of internal spores (sporangiospores). Serological tests, more particularly the immunodiffusion test using yeast phase or culture filtrate antigen (paracoccidioidin), are helpful tools in the laboratory diagnosis. The diagnostic value of skin tests is not fully established.

Antibiotic Sensitivity

Amphotericin B is considered the most effective agent in the management of paracoccidioidomycosis. However, its severe toxicity makes it hard to use this drug for long-term therapy. Success with ketoconazole and other azoles has been reported as well.

Sporotrichosis

Sporotrichosis is caused by *Sporothrix schenckii*, formerly known as *Sporotrichum schenckii*. Most mycologists list sporotrichosis among the causal agents of subcutaneous mycoses, but the authors chose to include this disease under the subtitle of systemic mycoses for three simple reasons. First, *S. schenckii* is a dimorphic fungus and dimorphic fungi are mostly associated with the systemic mycoses; second, *S. schenckii* does cause infections involving internal organs, such as the lungs; and third, systemic mycoses and subcutaneous mycoses are both deep-seated mycotic diseases and many systemic mycoses do exhibit subcutaneous or even mucocutaneous lesions. The natural reservoir of this fungus is soil, but it tends to colonize dead plant materials. For a time, sporotrichosis was called gardeners' disease because in a large number of cases the infection was traced to pricks by rose or barberry thorns. It now appears that in those cases, the thorns might have been contaminated with soil or vegetable matters harboring *S. schenckii*. Sporotrichosis is reported from all over the world. The disease is usually characterized by cutaneous and lymphatic lesions. Involvement of lungs has also been reported, albeit less frequently. The cutaneous lesions are often localized and develop on the torso, arms, neck, and occasionally face. The lymphatic involvement results from the dissemination of cutaneous infection. In most cases, the infections are acquired via traumatic implantation of the fungus due to a thorn or wood-splinter prick.

Laboratory Diagnosis

Homogenized biopsied tissues are inoculated on Sabouraud agar in duplicate and incubated at room temperature and at 35°C. At room temperature, the moist white colonies of *S. schenckii* develop within 1 week. The colonies develop darker pigmentation with age. The mycelia are thin and septate, and measure 1–2 μm in diameter. Conidia, measuring 2–3 × 3–6 μm, are hyaline, elliptical, and develop as a cluster at the tip of the mycelium (conidiophore) or laterally in singles on the side of the mycelia (Fig. 14.5). The yeast phase seen in cultures grown at 35°C or in

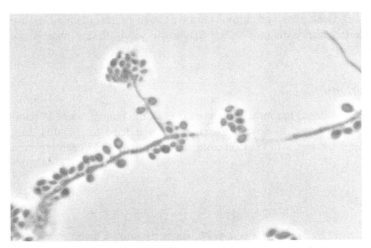

Figure 14.5.   Photomicrograph of the mycelial growth of *S. schenckii* showing typical conidial arrangement (lactophenol-cotton blue stain; courtesy: CDC).

biopsied tissue, *S. schenckii*, appears as spherical budding yeast-like cells measuring 8–10 μm in diameter. The yeast phase may also appear as cigar-shaped bodies measuring 1–3 × 3–10 μm in size.

Antibiotic Sensitivity

Most cases of cutaneous sporotrichosis respond well to Itraconazole although in the past oral administration of potassium iodide has been used successfully as well. Under certain circumstances, especially in disseminated cases, amphotericin B therapy is indicated.

DISEASES CAUSED BY YEAST-LIKE FUNGI

Cryptococcosis

Cryptococcosis, formerly known as Busse-Buschke's disease or European blastomycosis, is one of the most extensively studied mycotic diseases. The disease was first described by Busse and Buschke in 1894. The causal agent, *Cryptococcus neoformans*, has been isolated from natural sources from many parts of the world. Soil is the natural reservoir, but the fungus is more frequently isolated from old dried pigeon excreta, perhaps due to the high concentration of creatinine in avian droppings. *Cryptococcus neoformans* is one of a small group of microorganisms that can utilize creatinine as its sole source of nitrogen. Its isolation from various plant materials has been reported from all over the world, but no scientific explanation has been offered. Cryptococcosis is an airborne infection that is acquired via inhalation of the yeast. The lungs are the primary site of infection from where it spreads to the central nervous system, for which it seems to have a predilection. Pulmonary infections are mostly benign and asymptomatic and heal without leaving a recognizable scar, or with a radiologically discernible encapsulated lesion called cryptococcoma. Hematogenous dissemination leading to the involvement of skin, bones, abdominal viscera, and CNS may occur. Cryptococcosis in its most familiar form is characterized by CNS involvement. Symptoms include progressively severe frontal, temporal, or postorbital headache, vertigo, nausea, and vomiting. The lesions may involve meninges or cerebral tissues. Eye involvement leading to loss of vision is also noted in some cases. The fulminating form of infection is almost always fatal. Immune response and tissue reactions are minimal or absent altogether. Like almost all other infectious diseases, symptomatic cryptococcosis is most frequently seen in AIDS patients but can be seen in other immunocompromised patients as well. Those with underlying liver disease may also be predisposed to meningeal cryptococcal infection.

The tissue reaction in cryptococcosis is usually minimal. However, at histopathological examination, tissue reactions ranging from mucoid to occasionally granulomatous can be seen. In some instances, small cystoid foci located in the gray matter can be seen. A clear halo measuring 3–5 μm in thickness separating the yeast wall from the cytoplasm of the histiocytes are visible in sections stained with hematoxylin-eosin. In the rare cases of primary pulmonary infection, nodules measuring 1–6 cm in diameter and located near the pleural surface can be noted.

Laboratory Diagnosis

Clinical specimens, which in most cases are cerebrospinal fluid, are inoculated on Sabouraud agar and on Staib's bird-seed agar and incubated at 25°C. Colonies of *C. neoformans* are mucoid and white to cream-colored on Sabouraud agar and brownish to dark brownish on bird seed (*Guizotia abyssinica* seeds) agar. Microscopic examination of the culture should be done by mixing a small amount of the growth with India ink. *Cryptococcus neoformans* cells are spherical, measuring 4–20 µm in diameter and mostly accompanied by a mucopolysaccharide capsule of varying thickness (Fig. 14.6). However, not all *C. neoformans* strains show a capsule when grown in cultures. Several strains studied by Staib and Mishra (1977) showed no capsule when grown in culture but turned out to be highly virulent when injected into white mice. These strains showed a large capsule in animal tissue. In histopathological preparations, the fungus is best visualized with mucicarmine and PAS (periodic acid-Schiff) stains. No dependable serological tests are available.

Antibiotic Sensitivity

Varying degrees of success have been seen with fluconazole. In cases involving the CNS, a combination therapy using amphotericin B and 5-flucytosine are indicated for initial management in immunocompromised patients, with a subsequent switch

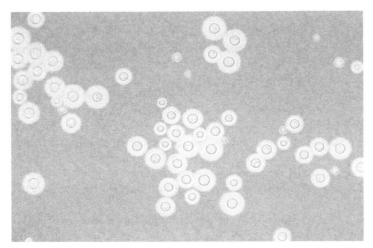

Figure 14.6.   *Cryptococcus neoformans* culture showing spherical cells with a large capsule (India ink preparation).

to oral fluconazole after an initial induction period with intravenous therapy in most instances.

Candidiasis

Candida infections are best known for thrush and vulvovaginitis, but systemic infections are also known to occur. Systemic candidiasis is an endogenous disease, mostly resulting from a hematogenous dissemination of the fungus. Cases involving endocarditis, septicemia, meninges, and the spleen, kidneys, and lungs have been frequently reported. Of all these, lung involvement or pulmonary candidiasis is almost always equivocal. Animal experiments using intravenous and intraperitoneal routes of inoculation have seldom shown lung involvement. Septicemia or candidemia is perhaps the most common form of systemic candidiasis. Species most frequently indicated include *Candida albicans, C. tropicalis, C. parapsilosis*, and *Torulopsis glabrata* (Fig. 14.7). Except for candidemia, which can occur in immune-competent individuals, all other forms of systemic candidiasis mostly occur in immunocompromised hosts.

Laboratory Diagnosis

Cultures of biopsied tissue are made on Sabouraud agar and incubated at 25°C and 35°C for up to 1 week. Media specially designed for blood culture are commercially available and useful. Species-specific differentiating characteristics have been discussed earlier. Biopsied tissues can be stained with hematoxylin-eosin or Gridley fungus stain. Immunodiffusion tests have been used but only with limited success.

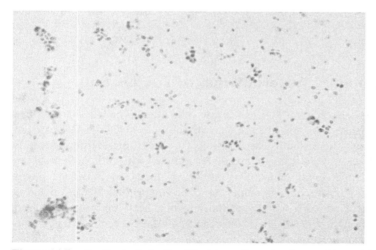

Figure 14.7.   Small spherical cells of *T. glabrata* (lactophenol-cotton blue stain).

Antibiotic Sensitivity

Systemic candidiasis, especially cases of endocarditis, meningitis, and candidemia, have traditionally necessitated amphotericin B infusion. However, a number of newer agents have been developed, including echinocandins and newer generation azoles, which have become the mainstay of treatment for candidemia alone with a removable focus. Amphotericin is still the preferred agent for endocarditis.

DISEASES CAUSED BY FILAMENTOUS FUNGI

Aspergillosis

Aspergillosis is an airborne infection and refers to a range of clinical conditions caused by *Aspergillus* spp., principally *A. fumigatus* and occasionally *A. niger, A. flavus*, and *A. terreus*. In 1975, Kwon-Chung proposed a new pathogenic species named *Aspergillus phialiseptus*. Between 1980 and 1985, Mishra et al. conducted in-depth studies on several similar strains isolated from the cases of aspergilloma and concluded that the so-called *Aspergillus phialiseptus* was actually *Aspergillus fumigatus*. The variations in the structure of the septa were probably a result of the host–microbe interactions. The "new" species name *Aspergillus phialiseptus* was subsequently dropped in favor of *Aspergillus fumigatus*. *Aspergillus* spp. are ubiquitous fungi commonly present in soil and easily airborne. *Aspergillus fumigatus* is a thermophilic species and usually associated with compost and other self-heated ecosystems. Cases of aspergillosis have been reported from all over the world. Most *Aspergillus* spp. produce toxins or toxic metabolites, including aflatoxins, aspergillic acid, and oxalic acid, which may complicate the clinical condition. Diseases associated with *Aspergillus* spp. include otomycosis, sinusitis, allergies and systemic infections. The following three major clinical forms of systemic aspergillosis are generally recognized.

Bronchopulmonary Aspergilloma, or Fungus Ball

Bronchopulmonary aspergilloma, sometimes also called fungus ball, is characterized by the growth of *Aspergillus* spp. in a pre-existing cavity as noted in some cases of cavitary tuberculosis. There is no tissue invasion. The "ball," measuring 1–5 cm, is formed by a compact growth of fungal mycelium admixed with cellular debris. The symptoms may include cough, hemoptysis (vomiting blood), and other symptoms related to pulmonary obstruction. In some cases, the fungus ball breaks up and is expectorated in pieces or as a whole and the infection heals up without clinical intervention. However, in some instances the condition is fatal if no clinical intervention, such as surgical resection, is made.

Invasive Aspergillosis

Invasive aspergillosis, though occasionally noted in immune competent individuals, is an opportunistic infection frequently associated with leukemia, cancer, diabetes, and AIDS. Lungs are the primary site of infection, but the disease may spread to other organs including kidneys and the CNS. The fungus invades and grows in the tissues. Pulmonary aspergillosis is characterized by granulomatous lesions with peripheral extension into lung parenchyma. The patient may expectorate mucopurulent sputum which may be blood-tinged. Kidney and CNS infections present classic symptoms associated with the involvement of those organs.

Allergic Bronchopulmonary Aspergillosis

Allergic bronchopulmonary aspergillosis (ABPA) is a syndrome that includes elements of infection and allergy. The pathophysiology involves both IgE and IgG immunoglobulins. The symptoms include episodic airway obstruction, fever, shifting pulmonary shadows, and expectoration of characteristic plugs in the sputum. This syndrome is known to occur all over the world, but its occurrence in the United States is seldom reported. Most cases of ABPA are reported from the United Kingdom and India. Cases of ABPA are treated with glucocorticoids to limit inflammation or a combination of glucocorticoids and Itraconazole is also used. Newer azoles that are used for invasive *Aspergillus* infections, such as voriconazole or posaconazole, have yet to be tested in clinical trials in comparison to itraconazole for ABPA.

Laboratory Diagnosis

In the case of bronchopulmonary aspergilloma, the diagnosis is usually aided by classic radiological appearance (intracavitary ball with an air crescent). Cultures are hard to obtain unless the fungus ball breaks up and the patient begins expectorating pieces. Surgically removed fungus balls can be cultured on Sabouraud agar supplemented with chloramphenicol. Serological tests can be helpful. Immunodiffusion tests generally reveal prominent precipitin bands.

In the case of invasive aspergillosis, cultures are helpful in some cases, especially those with lung or CNS involvement. In general, invasive aspergillosis is hard to diagnose. Cultures are not always successful. The role of serology is usually questionable, though conclusive immunodiffusion tests have been noted in some cases. Perhaps the best approach to a differential diagnosis is a combination of clinical, serological, and cultural findings.

In the case of ABPA, the laboratory diagnosis depends on a judicious consideration of clinical, radiological, cultural, and serological findings. Cultures are mostly positive if attempted during the episodes. Mycelium can be seen in the plugs by conducting a microscopic examination. Immunodiffusion tests are generally positive and stronger during the periods of the episodic obstruction.

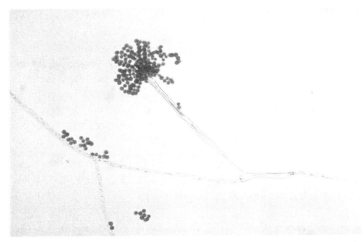

Figure 14.8.   Photomicrograph of *A. fumigatus* with a flask-shaped head and chain of conidia.

Aspergillus spp. grow fast on Sabouraud agar. The colonies in the case of *A. fumigatus* are smoky green, those of *A. flavus* are yellowish green, and those of *A. niger* are black. Mature mycelia develop characteristic spores borne on the sterigmata attached to globose or flask-shaped head on the conidiophores (Fig 14.8). The species are often identified on the basis of colony color, texture, and shape of the head.

Antibiotic Sensitivity and Therapeutic Management

In the case of bronchopulmonary aspergilloma, surgical removal of the fungus ball, followed by local application of amphotericin B or pimaricin (natamycin) solution, is usually practiced. Management of ABPA requires a judicious use of cortisone often with addition of Itraconazole, although newer azoles such as voriconazole are being used instead. The traditional choice of drug for invasive aspergillosis has been intravenous amphotericin B infusion, however, newer azoles such as voriconazole have demonstrated excellent efficacy with far less toxicity and the option for oral treatment. Echinocandins have been used as adjunctive treatment as well as second-line treatment for amphotericin B failures.

Zygomycosis

Zygomycosis, also known as mucormycosis, phycomycosis, and oomycosis, is caused by members of the division Phycomycota, also called zygomycetes. The number of species reportedly indicated in human infections is large, but the most frequently noted taxa belong to genera *Mucor, Rhizopus, Rhizomucor,* and *Absidia*.

Like aspergilli, these fungi are ubiquitous and commonly present in air and soil. The infections have been reported from all over the world. Zygomycosis is an opportunistic infection, frequently noted in patients with diabetes mellitus, and in other immune-compromised patients. The infection may involve the skin, lungs, paranasal sinuses, and gastrointestinal tract. Skin infections are generally localized and marked by distinct swelling. Pulmonary infections may present symptoms of bronchial or lobar pneumonia and gastrointestinal infections can cause blockage of the gastrointestinal tract. During the course of hematogenous dissemination, the fungus tends to invade major blood vessels and cause emboli or clot formation. The course of systemic zygomycosis is fast. Sever fulminating infections can develop in 2–10 days, resulting in the patient's death.

Emmons et al. have divided infections caused by this group of fungi into five subgroups, namely mucormycosis, entomophthoramycosis, oomycosis, chytridiomycosis, and trichomycosis. The causal agents in mucormycosis include *Mucor, Absidia, Rhizopus,* and *Cunninghamella,* which are often noted in fulminating lesions. The chief pathogens associated with entomophthoramycosis are members of the order Entomophthorales and mostly cause subcutaneous infections. Oomycosis, a heterogeneous group of mycoses, is caused by members of the class Oomycota and mostly affect animals. The term chytridiomycosis is used for the well-established disease entity rhinosporidiosis, characterized by nasal polyps. Causal agents of trichomycosis generally affect lower animals.

Laboratory Diagnosis

In the case of pulmonary zygomycosis, fresh morning sputum samples collected on three consecutive days are cultured on Sabouraud agar supplemented with chloramphenicol and incubated at 25°C and 35°C for up to 1 week. The biopsied tissues should be homogenized and processed in a similar manner. With a few exceptions, members of the division Phycomycota are aseptate. Hyphae are broad, measuring 10–30 μm in diameter. Asexual spores, called sporangiospores, are borne in a globose to subglobose structure called sporangium which is borne on a sporangiophore (Fig. 14.9). Specific identification is based on numerous factors including characteristics of conidiophores, presence or absence of rhizoids, and the arrangement of sporangia. The usefulness of serological tests has not been proven. In histopathological preparations stained with hematoxylin-eosin, hyphae of the causal agents of zygomycosis are easily seen.

Antibiotic Sensitivity

The course of infection is usually rapid in zygomycosis involving the CNS; therefore, many cases are diagnosed postmortem. Prognosis is poor in cases involving sinuses and the orbit and aggressive surgical debridement is required. Drug options are limited and a combination of amphotericin B and posaconazole are often used with varying success. In diabetics, blood sugar control must be a priority.

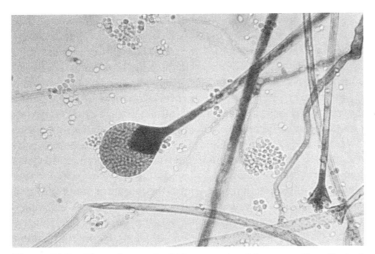

Figure 14.9.   Photomicrograph of *Mucor* spp. in a laboratory culture (Lactophenol cotton blue stain; courtesy: CDC).

Cerebral Chromomycosis

Cerebral chromomycosis, also known as cladosporiosis and cerebral dematiomycosis, is caused by a dematiaceous fungus *Cladosporium trichoides*, which is sometimes referred to as *Cladosporium carrionii, Cladosporium bantiana,* or more recently *Cladophialophora bantiana.* There has been some controversy as to whether cerebral chromomycosis should be considered as a separate entity or simply treated as a metastatic form of chromoblastomycosis, which is a common subcutaneous fungal infection. Emmons, a pioneer in medical mycology, has chosen to treat it separately from chromoblastomycosis because of the great differences in the histopathology of the lesions and, as he puts it, "the differences in the causality." In this context, the authors prefer to follow Emmons' school of thought.

The first case of cerebral chromomycosis in the United States was reported by Binford et al. in 1952. Since then nearly 50 cases have been reported from more than 20 countries. The clinical symptoms in cerebral chromomycosis are similar to other localized brain lesions. Most cases also present evidence of meningeal involvement. The disease is fatal if timely surgical intervention is not made. However, even after surgical removal of the lesions, long-term survival of patients is uncommon. Soil is the primary natural habitat of *Cladosporium trichoides* and infections are probably airborne.

Laboratory Diagnosis

Because there is no characteristic clinical syndrome, the differential diagnosis is based only on the laboratory findings. Isolation of the fungus in culture from blood or cerebrospinal fluid is rare. Histopathological and cultural examinations of the surgically removed tissues are of immense importance in the differential diagnosis. For isolation in culture, the homogenized or finely cut biopsied tissue is placed on Sabouraud dextrose agar supplemented with chloramphenicol and incubated at 35°C. To minimize the risk of contamination, slanted media in large test tubes is preferred over the traditional Petri dishes. The colonies are olive gray to olive brown with a velvety surface. At microscopic examination, brown septate conidiophores bearing a chain of conidia can be seen. The conidia are elliptical, pale brown, and measure $2-2.5 \times 4-7$ μm. Darkly pigmented mycelia and even chains of conidia can be seen in histopathological sections stained with Gridley stain or even in unstained sections. There are no serological tests.

Antibiotic Sensitivity

Surgical removal of the infected tissue is the only available course for the clinical management of cerebral chromomycosis. So far, the usefulness of amphotericin B and other antifungal agents has not been evaluated. As stated earlier, long-term survival of patients is rare.

DISEASES CAUSED BY MISCELLANEOUS FILAMENTOUS FUNGI

In the author's view, it is hard to draw a sharp line between pathogenic and non-pathogenic species. If the patient's immunity is severely impaired and the exposure

Figure 14.10.   Photomicrograph of *Penicillium marneffei*, isolated from a patient suffering from AIDS (lactophenol cotton blue stain; courtesy: Dr. Libero Ajello).

is heavy, almost any microorganism can create a disease-like condition. And fungi are no exception. In this era of stressful life styles and widespread use of cortico-steroids and other immunosuppressive drugs, several fungi that were once consid-ered harmless are now emerging as causal agents of serious infections in humans. Some of the examples include *Pneumocystis carinii* (now *P. jirovecii*), which com-monly infects the lungs of those with AIDS, *Penicillium marneffei* (Fig. 14.10), and many others that are beyond the scope of this concise manual.

Chapter 15

Unicellular Parasites

The term "parasite" is somewhat misleading. In medical microbiology it can mean any microorganism in/on human body in a "live at the cost of the host" relationship. However, in medical sciences, the term "parasite" usually means unicellular protozoa or multicellular metazoa (helminths) that cause diseases in humans. Protozoa are eukaryotes and belong to kingdom Protista. They are unicellular and devoid of a cell wall. The reproduction can be sexual or asexual. Classification of Protista or protozoa is complex and not without controversies. The International Society of Protistologists has divided them into 6 supergroups, which are further divided into 29 ranks. From a practical perspective, protozoa are divided into four groups based on their means of locomotion: Sarcodina (amoebas), Mastigophora (flagellates), Ciliophora (ciliates), and Sporozoa (stationary forms). However, from the perspective of pathogenic microbiology, the protozoa are divided into two groups.

- **Lumen-Dwelling Parasites:** Examples include *Entamoeba histolytica*, *Giardia lamblia*, *Cryptosporidium purvum*, and *Trichomonas vaginalis*.
- **Blood and Tissue Protozoa:** Examples include species of *Plasmodium*, *Leishmania*, and *Toxoplasma*.

LABORATORY METHODS IN PARASITOLOGY

For the sake of convenience, the laboratory methods commonly employed in medical parasitology are discussed here in general terms. Diagnosis is usually based on the direct examination of the clinical specimens, which may include feces, blood, or swabs from the affected regions. Serological tests, such as ELISA, agglutination, and fluorescence microscopy, are useful in some cases. Cultures possible only in a few cases are seldom useful in the diagnosis. Certain terms that will be used in this chapter are explained as follows.

- **Cyst:** Resting stage (resistant and dormant).
- **Oocyst:** Encysted fertilized female gametes in which sporozoites are formed.

A Concise Manual of Pathogenic Microbiology, First Edition. Saroj K. Mishra and Dipti Agrawal.
© 2013 Wiley-Blackwell. Published 2013 by John Wiley & Sons, Inc.

- **Sporozoite:** Sexual; spores (propagules) formed by fusion of male and female gametes and cell division resulting in haploid cells (cells with single set of chromosomes).
- **Trophozoite:** Feeding stage.

Collection of Clinical Specimens and Microscopic Examination

Fecal Specimens

Freshly collected samples should be processed within 2–3 hours. Preservatives, such as merthiolate-iodine-formalin (MIF) or sodium acetate-acetic acid (SA), should be used if required (especially during transportation). Specimens can be concentrated using suitable methods. For microscopic examination, a suspension of fecal matter is prepared in saline solution and a drop is placed on a slide, covered with coverslip, and examined under a microscope. Specimens smeared on a slide can be stained with Giemsa, Gomori's silver methanamine, or Trichrome stains and stored for permanent records.

Blood

Venous blood is generally used for the serological tests. For microscopic examination, a drop of blood can be obtained by puncturing the fingertip, heel of the foot, or earlobe. In certain cases, late-night samples are needed. Urine, genital fluid, sputum, and so on can be examined in the relevant cases.

In a deviation from the format used in earlier chapters, we have refrained from the use of the term "antibiotic sensitivity" with reference to parasitic diseases because most of the therapeutic agents used in the clinical management do not meet the classic concept of "antibiotics."

DISEASES CAUSED BY LUMEN-DWELLING PROTOZOA

Entamoeba histolytica (Amoeboid)

Entamoeba histolytica causes amebic dysentery worldwide. The actual estimate of people affected each year ranges from 100 to 400 million globally. The infection in most cases is benign and subclinical, and in some it may become chronic. In acute cases, the disease is characterized by abdominal discomfort, malaise, and frequent bowel movement. The fecal matter may contain mucous and occasionally blood. If not controlled in a timely manner, fever, dehydration, and electrolyte imbalance ensues. The infection is mostly food- and waterborne. The target tissue is the colon epithelium and the attachment is mediated by lectins and N-acetyl glucosamine conjugates.

Laboratory Diagnosis

Laboratory diagnosis is confirmed by the demonstration of characteristic trophozoites and cysts in the feces.

Effective Agents

Metronidazole (flagyl) is the drug of choice for the management of *E. histolytica* infections.

Giardia lamblia (Flagellate)

Giardia lamblia causes giardiasis, characterized by diarrhea with foul-smelling, mucous-laden stool, nausea, and cramps. Cases of giardiasis have been reported from all over the world. The prevalence in tropics may range from 2% to 30% of the general population. Nearly 50% of infected persons are asymptomatic. The disease can be chronic in IgA deficient subjects.

Laboratory Diagnosis

Demonstration of trophozoites or cysts in stool is a dependable diagnostic tool. *Giardia lamblia* cysts are oval in shape and measure $8–12 \times 7–10$ μm.

Effective Agents

Giardia is effectively treated by either tinidazole or metronidazole in most patients.

Trichomonas vaginalis (Flagellate)

Trichomonas vaginalis causes trichomoniasis or vaginitis characterized by itching, burning sensation, dysuria, and thick yellow discharges. The diseases can be sexually transmitted, though males are usually asymptomatic. This protozoan mostly grows in the female urogenital tract by establishing colonies of trophozoites on the surfaces of the vaginal epithelium. Its prevalence in female population ranges from 10% to 60%. The disease is mostly transmitted via sexual intercourse. Transmission may also occur from female to female through sharing of intimate garments or washcloths. *Trichomonas vaginalis* cannot survive more than a few hours under dry conditions.

Laboratory Diagnosis

Laboratory diagnosis can be made promptly by a microscopic demonstration of motile trophozoites in the vaginal fluid (Fig. 15.1). Cysts are seldom seen. Several commercial kits and PCR method are also available.

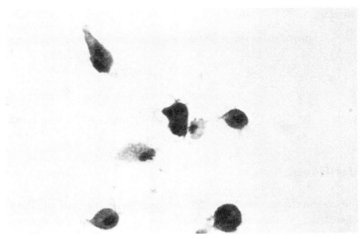

Figure 15.1.   A photomicrograph depicting *Trichomonas vaginalis* cells (Giemsa stain).

Effective Agents

Metronidazole is the most useful agent for the management of trichomoniasis.

Cryptosporidium parvum (Sporozoa)

Cryptosporidium parvum causes cryptosporidiosis, characterized by watery diarrhea, cramping, abdominal pain, and weight loss. Its incubation period is 2–3 weeks. This protozoan is quite resistant to chlorine and ozone and it is frequently isolated from water and several animal species. Cryptosporidiosis can be a waterborne or a zoonotic disease. Person-to-person transmission may also occur. In past, *C. parvum* was never taken seriously as a pathogen until the "emergence" of AIDS. Its prevalence rate in AIDS cases is estimated to be 10%–20%.

Laboratory Diagnosis

Microscopic demonstration of oocysts is a dependable tool for the laboratory diagnosis.

Effective Agents

No satisfactory chemotherapeutic agent is currently available for the clinical management of cryptosporidiosis.

BLOOD- AND TISSUE-DWELLING PROTOZOA

Plasmodium Species (Malarial Parasite)

Malaria is perhaps the most devastating disease in the tropics and subtropics. According to World Health Organization, globally more than 250 million people suffered from malaria in 2009. The annual worldwide mortality was in excess of one million, mostly in sub-Saharan Africa. Considering the fact that in poor and underdeveloped countries a large number of cases go unreported, the actual number may be much higher. Cases of malaria have been reported from many different parts of the world, probably brought in by the travelers. In the United States, nearly 1,000 cases are noted each year. The disease is characterized by mild headache, nausea, and chills lasting 10–15 minutes, followed by high fever for 2–20 hours. The cycle may be repeated at 24-, 48-, or 72-hour intervals. Untreated cases are usually fatal. Malaria is a zoonotic disease. Female mosquitoes, *Anopheles*, are the alternate hosts and vectors of the disease. The incubation period is variable. The lifecycle of malarial parasite is illustrated in Figure 15.2.

Five species of the genus *Plasmodium* are known to cause malaria in humans. These are *P. falciparum, P. ovale, P. vivax, P. malariae,* and *P. knowlesi. Plasmodium knowlesi* was first described in Malaysia in 2004, but documented infections have been apparent since 1996. It may look like *P. malariae* under microscopic examination, but can have a high parasite density and causes a wide spectrum of illnesses. Infections due to *P. falciparum* are most severe. The pathogen is injected into the human body through the bite of female *Anopheles* mosquito. From the blood stream, the sporozoites enter the liver, produce merozoites, and are released back into the blood stream. The merozoites invade the erythrocytes and form trophozoites, which in turn produce schizonts (Fig. 15.3). The schizonts upon maturity produce merozoites leading to the lysis of the erythrocytes. The erythrocytic stage is repeated every 48–72 hours.

Laboratory Diagnosis

In the endemic areas, experienced physicians diagnose malaria mostly on the basis of clinical symptoms. A laboratory diagnosis can be established by the demonstration of species specific trophozoites (10–15 μm long) in the Giemsa-stained blood smears.

Effective Agents

Infections by *P. vivax, P. malariae,* and *P. ovale* are usually mild and treatable with chloroquine and primaquine. Infections by *P. knowlesi* are also treated with chloroquine. However, *P. falciparum* are resistant to chloroquine, but respond to combined therapy with mefloquine and doxycycline. A prophylactic preparation containing chloroquine, taken once a week, is often recommended for travelers planning to visit

Malaria
(Plasmodium spp.)

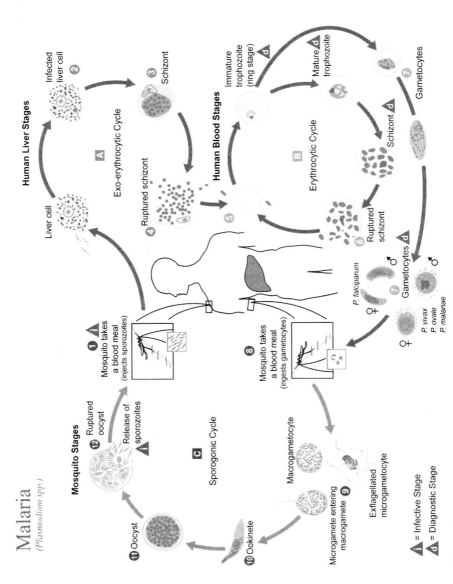

Figure 15.2. Life cycle of the malarial parasite (source: CDC).

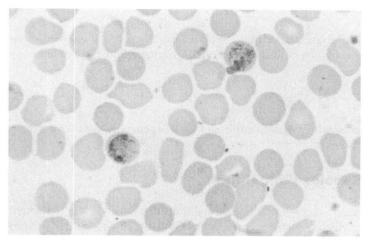

Figure 15.3.   A blood smear stained with Giemsa stain depicting two red blood cells with *Plasmodium malariae* schizonts. See color insert.

endemic areas. So far, there is no effective vaccine in mass circulation, but success has been recently reported with the development of an effective vaccine.

Toxoplasma gondii

Toxoplasma gondii causes toxoplasmosis, a disease of worldwide occurrence. The prevalence rate is estimated to be 20%–60%. Cases of toxoplasmosis are noted in the United States with a regular frequency. The disease is mostly acquired through ingestion of raw or undercooked meat from cattle, sheep, pigs, or chickens. In some cases, infection can occur via consumption of food or fruits contaminated with the oocysts found in cat feces. Pregnant women, if infected with *T. gondii*, can pass the infection to the baby through the placenta. The disease is characterized by inflammatory lesions involving the reticuloendothelial system (liver, spleen, and lymph nodes), lungs, heart, brain, and eyes. Symptoms may include fever, fatigue, and enlarged liver or spleen. Encephalitis, hepatitis, or pneumonia may develop in advanced cases.

Two distinct life cycles of the parasite have been noted: enteric and asexual cycles. In the enteric cycle, *T. gondii* grows in the intestinal epithelium of cats and other members of the feline family, which are considered to be the main carriers. The oocysts, produced as a result of sporogeny, are excreted in the feces, and are capable of surviving in the external environment for more than a year. The asexual cycle begins in animals such as cattle, pigs, and chicken, following ingestion of materials contaminated with the oocysts. The pathogen can survive as oocysts in these alternate hosts and infect humans upon consumption of improperly cooked meat.

Diagnosis

The most dependable diagnostic tool for the toxoplasmosis is based on the demonstration of specific antibodies using ELISA or IFA (immunofluorescence antibody assay).

Effective Agents

A combination therapy with pyrimethamine and sulfa is generally used for the management of toxoplasmosis.

Leishmania Species (Flagellate)

Leishmania spp. cause a number of clinical conditions that are collectively referred to as leishmaniasis, which occur all over the world. Globally, about 2 million persons are believed to suffer from leishmaniasis each year. Of these, more than 50,000 die annually. The primary reservoir of *Leishmania* spp. are rodents and certain members of the canine family. Female sand flies, mostly those belonging to the genera *Lutzomyia* and *Phlebotomus*, are the carriers and the chief vectors of the disease. During the course of their blood meal, the flies inject promastigotes (a stage in which a single flagellum arises from a kinetoplast located at the anterior end of the parasite) into the human body. The promastigotes multiply within the macrophages and form amastigotes, which are nonmotile. The host cells are destroyed in the process. Three distinct clinical manifestations, with drastically different outcomes, are noted in leishmaniasis. These are summarized below.

Cutaneous Leishmaniasis

Cutaneous lesions are usually caused by *L. tropica*, which occurs mostly in the deserts of Arabia and Northern Africa. *Leishmania mexicana*, another species known to cause skin lesions, is found in the Yucatan peninsula and neighboring parts of Mexico, and in the adjacent regions of Texas. The disease is characterized by small, red papules, usually at the site of the bite by the sand fly. The lesions that eventually become crusted ulcers are often seen on the face and ears. The infection is mostly self-healing.

Mucocutaneous Leishmaniasis

Leishmania brasiliensis is usually associated with mucocutaneous lesions on the nose, throat, mouth, and skin. The disease is mostly confined to Latin America, and it often results in extensive scarring and disfigurement.

Visceral Leishmaniasis

The causal agent of the visceral leishmanias, or Kala-azar (Sanskrit: *Kalajwar*) is *L. donovani* (Fig. 15.4), which is endemic in Northern China, Eastern India,

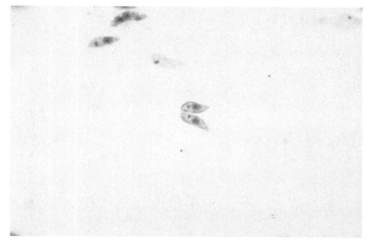

Figure 15.4.   *Leishmania donovani*, leptomonad forms (courtesy: CDC). See color insert.

Mediterranean countries, North Africa, and parts of Latin America. Its incubation period may range from a few days to several years. During the early phase of infection, symptoms are mild or absent. *Leishmania donovani* has a predilection for the reticuloendothelial system, especially for the macrophages found in the spleen, liver, bone marrow, and lymph nodes. Symptoms include intermittent fever and enlargement of the liver and spleen. Extensive growth of the parasite in the reticuloendothelial system destroys the affected organ. Among the famous victims of visceral leishmaniasis was Akbar the Great, who ruled India in the 16th century.

Laboratory Diagnosis

Microscopic demonstration of intracellular amastigotes in Giemsa-stained blood smears is helpful for the diagnosis of cutaneous leishmaniasis. In the case of mucocutaneous leishmaniasis, the diagnosis is often based on clinical history and Montenegro test. Laboratory diagnosis of visceral leishmaniasis can be challenging. It can be achieved by demonstrating promastigotes in tissue or culturing it on selective media including Schneider's medium. Serological tests, such as ELISA and IFA, are also dependable and easily performed.

Effective Agents

Most cases of leishmaniasis are effectively treated with sodium stibogluconate, available through the CDC.

Chapter 16

Multicellular Parasites

Multicellular parasites (metazoan) possess specialized organ systems (digestive, excretory, locomotive, etc.). The adults show a wide variation in their sizes, ranging from a few millimeters to several meters in length. They are commonly called "helminths" or "worms." The list of diseases caused by multicellular parasites is long. As in the case of bacterial, fungal, and protozoan diseases, only some typical and most commonly seen multicellular parasites are discussed here. Sizes of the mature adults and diagnostic eggs are summarized in Table 16.1.

Helminths can be divided into two major groups:

- **Roundworm (phylum Nematoda)**: Most roundworms have a complete digestive system, and there are separate male and female individuals.
- **Flatworm (Platyhelminths)**: Most flatworms have a primitive digestive system devoid of anal opening, and they are hermaphrodites.

Flatworms are further divided into:

- **Tapeworms (Cestoda)**
- **Flukeworms (Trematoda)**

Tapeworms absorb nutrients directly. They have a series of individual segments called proglottids. Each segment contains a separate set of male and female organs.

A number of helminths have multiple hosts. For the sake of convenience, pathogenic microbiologists prefer to divide helminths into two groups:

- Lumen-dwelling helminths
- Blood- and tissue-dwelling helminths

LUMEN-DWELLING HELMINTHS

Pinworm (*Enterobius vermicularis*)

Cases of pinworm infection occur all over the world including the United States. Children between the ages of 5 and 10 years are most frequently affected. Infection

A Concise Manual of Pathogenic Microbiology, First Edition. Saroj K. Mishra and Dipti Agrawal.
© 2013 Wiley-Blackwell. Published 2013 by John Wiley & Sons, Inc.

Table 16.1   Comparison of Some Multicellular Parasites with Reference to the Size of Mature Adults and the Size of Diagnostic Eggs

Species	Disease/ Common name	Size of the adults	Size of the diagnostic eggs
Enterobius vermicularis	Pinworm	Males: 2–5 mm Females: 8–13 mm	20–32 × 50–90 μm
Ascaris lumbricoides	Large Roundworm	Males: 15–31 cm Females: 20–35 cm	45–60 μm
Taenia saginata	Beef tapeworm	2–7 m	30–40 μm
Taenia solium	Pork tapeworm	5–10 m	30–40 μm
Schistosoma spp.	Schistosomiasis	7–20 mm	Variable*
Paragonimus westermani	Lung fluke	6 mm × 10 mm	55–90 μm

* *S. japonicum:* 55–85 μm in length, 40–60 μm in width; *S. mansoni:* 114–180 μm in length, 45–75 μm in width; *S. haematobium:* 112–170 μm in length, 40–70 μm in width.

is acquired by ingesting food contaminated with larvae, or direct from the hands of infected child who has touched the affected area or handled contaminated clothing. Symptoms include nocturnal perianal and perineal pruritus. At night the females come out through the anus and deposit 4,000 to 16,000 eggs on the perianal skin that may survive up to 10 days. The larvae may hatch in the perianal folds and then migrate into the anus and mature as adults in the intestine. This process of self-infection is also called retrofection.

Laboratory Diagnosis

A dependable method is to conduct microscopic examination of the swabs or sticky paddles taken from the anal folds (Fig. 16.1). The specimen should be collected immediately after the child wakes up. Pinworm eggs measure 50–60 × 20–30 μm in size.

Effective Agents

Mebendazole is the most frequently used drug for the treatment of pinworm infection.

Large Roundworm (*Ascaris lumbricoides*)

Ascaris infections occur worldwide, but most commonly in the tropical countries where it causes about 20,000 deaths each year. Most cases are subclinical and the actual number of infected people may run into millions. Acute infections are relatively uncommon and affect mostly young children younger than the age of 10.

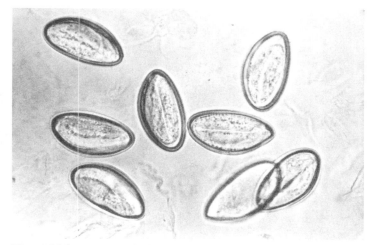

Figure 16.1.   Eggs of *Enterobius vermicularis* as captured on an adhesive tape. (courtesy: CDC).

Symptoms in acute cases include abdominal obstruction due to massive parasitic growth in the intestine. Infection is acquired through ingestion of helminthic eggs in raw food. The eggs hatch in the duodenum; larvae penetrate the intestinal wall, enter into blood circulation, and reach the lungs, where fourth-stage larvae are produced. The larvae then migrate to the pharynx and swallowed and reach the intestine where larvae mature into adults. Ascaris eggs are quite resistant to environmental conditions and capable of surviving in nature for more than 10 years.

Laboratory Diagnosis

The laboratory diagnosis is mostly based on the demonstration of fertilized (45–60 µm) or unfertilized (40–90 µm) eggs in the fecal matter. Occasionally, adult helminths measuring 15–31 cm (males) or 20–35 cm (females) can be seen in the feces and identified on the basis of their characteristic morphology.

Effective Agents

Mebendazole is most commonly used drug for the management of *Ascaris* infection.

Beef Tapeworm (*Taenia saginata*) and Pork Tapeworm (*Taenia solium*)

Infections involving *T. saginata* or *T. solium* occur worldwide, but more commonly in the developing countries including parts of Africa, South America, and even

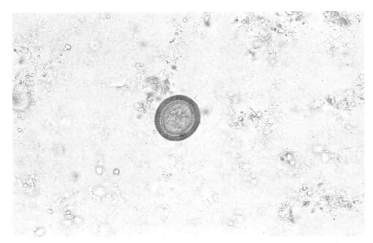

Figure 16.2.   An egg of *T. solium* under direct microscopic examination. Eggs of *T. solium* and *T. saginata* exhibit a high degree of similarity (source: CDC).

Russia. The disease is mostly associated with the consumption of undercooked or raw meat. The infection is acquired through ingestion of cysticercus stage of the larvae, found in the muscles of infected animals, also called intermediate hosts. *Taenia saginata* is generally associated with cattle and the *T. solium* is found in pigs. Clinical manifestations include abdominal discomfort. About 2 pounds of proglottids are produced each year, which may lead to severe malnutrition. In rare instances, invasion of the intestinal wall can occur and even the brain may be affected.

Laboratory Diagnosis

Laboratory diagnosis is often based on the demonstration of eggs (30 to 40 µm) or the characteristic proglottids in the fecal matter (Fig. 16.2). Serological tests using the immunoblot technique are also useful. Adult *T. solium* may measure 2–7 m and *T. saginata* 5–10 m in length.

Effective Agent

The drugs of choice in both kinds of tapeworm infections are praziquantel and niclosamide.

Schistosomiasis

The infection may pass through several phases. In the first phase, involvement of lungs leading to the pulmonary hemorrhage may be accompanied by cough and hemoptysis. During the secondary phase, the pathogen migrates to the mesentery vein leading to the development of serum sickness. At this stage, in some cases, fever, abdominal pain, and diarrhea may ensue. However, gross symptoms may differ from species to species. The causal agents of schistosomiasis differ from region to region. *Schistosoma japonicum* infections, mostly characterized by dysentery and enlargement of the liver and spleen, are generally noted in Japan, China, and the Philippines. Infection by another species, *Schistosoma mansoni,* mostly associated with inflammation of large intestine and liver, is common in South America, Africa, and Caribbean countries. Infections caused by a third species, *S. haematobium,* are mostly noted in North Africa, especially in the Nile Valley, and occasionally in the Middle East and India. *Schistosoma haematobium* infections mostly involve the urinary tract, causing cystitis and hematuria. The estimate of the number of people infected at any time runs into millions. Snails are the only alternate host. Infection occurs through penetration of exposed skin by infective cercaria (a free-swimming trematode larva that emerges from its host snail).

Laboratory Diagnosis

The laboratory diagnosis is mostly based on the demonstration of characteristic eggs in feces, biopsied tissue, or urine. *Schistosoma japonicum* eggs measure 55–85 µm, those of *S. haematobium* 112–170 µm, and the eggs of *S. mansoni* measure 114–180 µm. Serological tests including ELISA and immunoblot techniques are helpful. Adult males are 7–20 mm long and the females are slightly larger. Their life span is about 25–30 years. Males are generally seen in the perpetual embrace of the females.

Effective Agents

Clinical management of schistosomiasis is generally achieved through Praziquantel.

Hookworms

Several hundred million persons are believed to be infected with hookworm. Epidemiologically, hookworms are divided into two groups. The old world hookworm, caused by *Ancylostoma duodenale,* mostly occurs in the tropics and has a lifespan of about 1 year. The new world hookworm, caused by *Necator americanus,* is usually seen in the Americas. Infection is acquired through penetration of skin exposed to soil contaminated with the larvae. This disease is more common in rural areas. The life cycle of the parasite can be divided into three phases, each with a distinct set of symptoms. The first of the three phases, the cutaneous phase, results from the penetration of skin by larvae, and is characterized by dermatitis, pruritus, and erythema. The second phase, called the pulmonary phase, ensues following the migration of larvae into the lungs. No remarkable clinical symptoms are associated with this phase. The third phase, the intestinal phase, develops when the pathogen attaches to the mucosa of the small intestine. During this phase, on an average, each adult sucks 0.2 mL of blood each day. Considering the fact that several hundred adult worms can be found in the intestine of infected persons, a substantial loss of blood may occur, resulting in severe drop in the hemoglobin and leading to serious consequences.

Laboratory Diagnosis

The laboratory diagnosis is easily made through the demonstration of unembryonated eggs in feces. The eggs measure 40–60 µm. The adult males of *A. duodenale* measure 0.5 cm in width and about 1 cm in length. The females are slightly bigger. Adult *N. americanus* are somewhat smaller.

Effective Agents

The drug of choice for the clinical management of hookworm infection is mebendazole.

Lung Fluke (*Paragonimus westermani*)

Lung fluke can have several clinical manifestations. In the case of the lung involvement, the disease is characterized by chronic bronchitis with productive cough, mostly in the mornings. The condition is occasionally mistaken for chronic pulmonary fungal infection. At least in one case that the authors are familiar with, the lung fluke was misdiagnosed as a case of pulmonary candidiasis. Symptoms in the case of abdominal infection mostly include mild abdominal pain, which is occasionally accompanied by bloody diarrhea. Cerebral lesions are also noted in some cases. The lung fluke is mostly noted in Southeast Asia and parts of Africa and Latin America. The infection occurs following the ingestion of insufficiently cooked or uncooked food containing certain kinds of freshwater crabs or crayfish that are generally infested with *P. westermani*. Once in the intestine, the encysted stage of the worm called metacercaria is activated and enters the peritoneal cavity after penetrating through the intestinal wall. From peritoneum, it penetrates the diaphragm and reaches the lung parenchyma where it will develop into an adult and lay eggs. The eggs are eventually released from the lungs and expectorated in the sputum. The mature adults measure about 1 cm in length and 0.6 cm in width.

Laboratory Diagnosis

The laboratory diagnosis is mostly based on the demonstration of typical eggs in sputum or feces (Fig. 16.3). The eggs measure 55 µm in width and 90 µm in length. Serological tests can be useful, but they are not readily available.

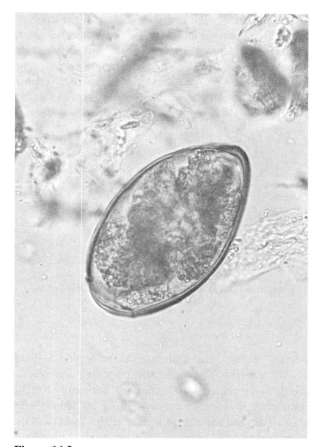

Figure 16.3.   *Paragonimus westermani* egg in an unstained, formalin-preserved fecal sample.

Effective Agents

Praziquantel is the most frequently prescribed chemotherapeutic agent for the management of the lung fluke.

BLOOD- AND TISSUE-DWELLING HELMINTHS

Trichinosis (*Trichinella spiralis*)

Trichinosis, essentially a muscle infection, presents a wide range of clinical symptoms. Most cases are asymptomatic, but acute infections may develop in some. The symptoms correlate with the specific phase of the infection. In the first phase, following ingestion of undercooked pork infested with larval cysts of *T. spiralis*, the symptoms may include abdominal pain, nausea, and diarrhea. The second phase begins following penetration of the intestinal wall and the subsequent presence of the worm in the circulatory system. Symptoms at this stage include headache, fever, muscle pain, and edema around the eyes. During the third phase, larvae penetrate the muscles, causing muscle pain. Pulmonary and neurologic symptoms may occur in some cases. Strains of *T. spiralis* are known to produce toxic metabolites. Trichinosis cases have been noted in many parts of the world. Many animals in arctic regions naturally suffer from this disease. Other animals known to harbor *T. spiralis* infection include dogs, cats, rats, and pigs.

Laboratory Diagnosis

Laboratory diagnosis is mostly based on the demonstration of encysted larvae in muscle biopsy. Serological tests, such as ELISA and bentonite flocculation test, are also helpful.

Effective Agent

Mebendazole is most frequently used in the clinical management of trichinosis.

Filariasis

The terms "filariasis" or "philariasis," or just "filarial disease," actually refer to three distinct sets of clinical conditions, namely, lymphatic, subcutaneous, and serous cavity filariasis. Lymphatic filariasis is caused by *Wuchereria bancrofti, Brugia malayi,* and *B. timori.* The subcutaneous form is mostly caused by *Mansonella streptocerca* and *Onchocerca volvulus,* and the causal agents of the serous cavity filariasis are often *Mansonella perstans* and *M. ozzardi.* All three clinical forms are zoonotic infections in which humans are the primary host, and the transmitting vectors are mosquitoes or black flies. According to some estimates, roughly 150 million people in South-Southeast Asia, North Africa, and Central and South America are infected with filariasis.

Of the three clinical forms, lymphatic filariasis caused by *W. bancrofti* has drawn the greatest attention. The infection is characterized by edema with thickening of the skin and underlying tissues, especially in the legs, arms, and scrotum. The condition may lead to a massive swelling of the lower legs, which is commonly called elephantiasis. The scrotum involvement, popularly known as hydrocele, is characterized by excessive enlargement of the scrotum, which may occasionally reach to about 10 inches in diameter. Most symptoms result from the cumulative damage caused by the worms, immune response, and accumulation of dead worms. Inguinal lymph nodes can be inflamed, swollen, red, and painful.

The infection begins with the introduction of filarial larvae into human skin following a bite by female mosquitoes, usually belonging to the species *Culex pipiens* or *Aides aegypti.* The larvae, measuring 1.4–2 mm long, migrate to peripheral lymphatic vessels mostly in the lower body where they develop into adult

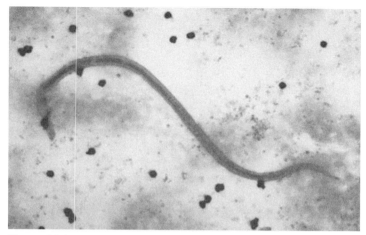

Figure 16.4. Photomicrograph of a *Wuchereria bancrofti* microfilaria in a thick blood smear using Giemsa stain (source: CDC). See color insert.

worms. The adult worms, which can live for 5–15 years, mate and produce a large number of elongated embryos called microfilaria, which circulate in the blood. Humans are the primary host and mosquitoes serve as the alternate host. The World Health Organization (WHO) is currently trying to eradicate filariasis by eradicating the parasite from humans. An ambitious program aimed at breaking its life cycle by treating the population at risk in the endemic areas with a once-a-year oral dose of mebendazole and ivermectin combined is currently under way.

Laboratory Diagnosis

The laboratory diagnosis is typically based on the microscopic demonstration of microfilariae in Giemsa-stained smears prepared from peripheral blood (Fig. 16.4). The microfilariae measure 250–300 μm in length. Blood sample should be collected during the night, preferably after 8 P.M. It is worth noting that the microfilariae are not always present in the peripheral blood. PCR tests have also been useful in the diagnosis of filariasis.

Effective Agents

The most commonly used drugs for the treatment of filariasis are ivermectin and mebendazole. The two drugs can also be used in combination. Both kill microfilariae, but have no effect on the adult worms. A combination of mebendazole and diethylcarbamazine is also used by some physicians.

Chapter 17

Viruses and Prions

Conservative microbiologists do not consider viruses as living organisms. They prefer to treat them as a bridge between the living and the nonliving. Virology is a highly advanced branch of science and many advanced centers of learning have a separate department of virology, independent of microbiology. Numerous excellent textbooks and reference books are available in virology. In this chapter, only the very basic concept is summarized. Viruses are obligate parasites and with a few exceptions, they cannot survive outside the host cells over an extended period of time.

Structurally, viruses are simple. The core structure of a virus is made of one nucleic acid, DNA or RNA (seldom both), surrounded by a protein coat called capsid. Some viruses (mostly animal viruses) are enclosed in an envelope that resembles cytoplasmic membrane. Viruses without an envelope or the outer membrane are called naked viruses. Viruses may have a helical shape, icosahedral (polyhedral) shape, or a combination of the two (these types are called the complex viruses). They are submicroscopic and generally invisible through bright field microscopy. Their size ranges from 20 to 800 nm, but most are less than 100 nm in size. Their genome is rather small and ranges from 7 kilobases (kb) in some RNA viruses to about 200 kb in DNA viruses, such as pox and herpes viruses.

Several factors are taken into account in the classification of viruses. These include types of nucleic acid, number of strands, composition of capsid, shape, and host. Also taken into account in the case of RNA viruses is whether the RNA is a (+) sense or (−) sense. A (+) sense virus genome can function as mRNA, which is capable of infecting a susceptible cell on its own, without presence of a protein coat. In contrast, the (−) sense RNA genome upon entering the host cell has to be copied to form mRNA. A simple and practical classification will be to divide viruses into four major groups.

- Double-stranded DNA (ds-DNA) viruses
- Single-stranded DNA (ss-DNA) viruses

A Concise Manual of Pathogenic Microbiology, First Edition. Saroj K. Mishra and Dipti Agrawal.

- Double-stranded RNA (ds-RNA) viruses
- Single-stranded RNA (ss-RNA) viruses

Further groupings are based on the presence of envelope, size and shape of the capsid, and molecular considerations.

LABORATORY DIAGNOSIS

Common viral diseases, such as influenza, herpes, and the common cold, are often diagnosed on the basis of clinical symptoms. Laboratory diagnosis may be required in others. Serological tests including ELISA, immune blot, and molecular techniques such as PCR are widely available for several viral diseases. Tissue culture is possible in many cases, but it is less commonly attempted because of the high cost and longer time required for a definitive result. Appropriate clinical specimens, such as venous blood or swab samples, are collected aseptically and promptly sent to the laboratory.

Viruses are a major cause of mortality and morbidity worldwide. Some of the deadliest diseases, such as smallpox, Ebola, AIDS, and rabies, are caused by viruses. Fortunately, the advent of effective vaccines has played a remarkable role in the eradication of smallpox and control of several other deadly viral diseases. For the sake of brevity, we will briefly address common viral diseases with a brief discussion on a few. Listed below are some of the families of important viruses that cause diseases in humans with a succinct comment on these diseases.

DOUBLE-STRANDED DNA VIRUSES

Papavoviridae (Papilloma Viruses)

Members of Papavoviridae are icosahedral naked viruses that multiply in the nucleus. An important member of this family, human papillomavirus (HPV), causes papilloma and warts in humans. The virion is comprised of six early proteins, termed E1, E2, E3, E4, E6, and E7, and two late proteins, L1 and L2. The E6 and E7 proteins are mostly associated with cancer. They are known to inactivate tumor-suppressing proteins p53 and pRb. Of the more than 100 types of HPV known, Types 16, 18, 31, 33, 35, 39, 45, 51, 52, 56, 58, and 59 are believed to be high risk viruses responsible for sexually transmitted diseases. HPV type 16 and HPV type 18 are most frequently associated with cervical cancer and HPV type 6 and HPV type 11 are most commonly associated with genital warts. The virus generally infects keratinocytes of the skin or mucous membrane. In most cases, the infections are sexually transmitted from person to person either through sexual intercourse or oral sex or through contact with skin around the genitals of infected persons.

The HPV viruses are generally associated with four types of clinical manifestations: skin warts, respiratory infection (recurrent papillomatosis), genital warts, and cervical infections. Skin warts are noncancerous skin growths, including common warts on hands, feet, and inner thighs, which may be characterized by cauliflower-like appearance. Another type of skin wart, the plantar warts, are found on the soles of the feet and tend to grow inward, making it painful to walk. The respiratory tract infection mostly associated with HPV type 6 and 11 results in the formation of warts on the larynx and other areas. The respiratory warts interfere with breathing and require repetitive surgery. Genital warts or venereal warts also known as condylo-mata acuminata are easily recognized symptoms of genital HPV infection. At least 40 HPV types have been found in association with genital warts, but HPV types 6 and 11 account for 90% of cases. Genital warts are mostly seen in the vulvar areas, vagina, and on or around anus. The cervical infections caused by types 16 and 18 are mostly associated with neoplasia involving cervical, vulvar, penile, or anal intraepithelial tissues. Another important type of HPV-related cancer is oropharyn-geal cancer. In most cases, HPV infections are cleared by the immune system. Clinically manifested infections develop only in a small percentage of persons. Of the four types, HPV infections associated with cancer are most serious. More than 5% of all cancer cases are believed to be due to HPV. More than 80% of cervical cancer cases are due to HPV infection. The incidence rate of HPV-associated oro-pharyngeal cancer in the United States is almost equal to the incidence rate of oropharyngeal cancer due to tobacco usage. In 2008, approximately 12,000 cases of cervical cancer were diagnosed in the United States with nearly 4,000 fatalities.

HPV-related skin and genital warts are often diagnosed on the basis of clinical symptoms. Regular Papanicolaou (Pap) smear examination is most effective in detecting cervical cancer in its early stages. A recently introduced hybrid-capture test can be used along with Pap smear testing. A confirmatory test may require colposcopy and histopathological examination of the biopsied tissue. Currently, two vaccines are available for preventing cervical cancer. These include Gardasil (Merck) and Cervarix (GlaxoSmithKline), which are effective only if given to noninfected females. They protect against the cancer associated HPV types 16 and 18. The vac-cines are usually given in three doses over a 6-month period. Duration of the effec-tiveness of vaccination is still not proven. These vaccines have been in use only during the past few years. Use of condoms is not fully protective because areas around the genitals, including the inner thighs of infected females, may harbor HPV. Condoms play no prophylactic role in oral sex; finger-to-genital contact can also be risky. Certain topical microbicides and sexual lubricants are believed to block HPV transmission if applied to the genitals prior to sexual contact. However, their useful-ness is yet to be clinically proven. There is no effective treatment for HPV infections. Cryotherapy or freezing appear to be most effective therapeutic measure in prevent-ing the infection from developing into cervical cancer.

Adenoviridae (Adenoviruses)

Adenoviruses are naked icosahedral viruses that multiply in the nucleus. They are known to have several serotypes that differ from each other with reference to their molecular mass and capsid protein. Nearly 46 serotypes, divided into six subgroups named A through F, are associated with human diseases. Infections mostly involve the respiratory tract and may present pertussis or pneumonia-like symptoms, often accompanied by fever. Other clinical conditions associated with the adenoviruses include hemorrhagic cystitis, gastroenteritis, and conjunctivitis (pink eye). Children are generally at greater risk.

Herpesviridae (Herpes Viruses)

Herpesviridae represents a somewhat heterogeneous group of viruses that are enveloped, have an icosahedral capsid, and multiply in the nucleus. They are divided into three subfamilies:

Alphaherpesvirinae

Members of the subfamily Alphaherpesvirinae includes human herpes virus 1 (HHV-1) also called Herpes Simplex Virus 1 (HSV-1), HHV-2 (HSV-2), and HHV-3, usually called Varicella-zoster (chicken pox) virus. Members of this family target mucoepithelial cells. During their latency period, they tend to reside in the neurons. Some of the differences between HSV-1 and HSV-2 are summarized in Table 17.1.

Varicella-zoster causes chicken pox, which mostly affects children younger than 10 years of age, though cases of infections in adults are not rare. A chicken pox vaccine has been developed, but its usefulness still awaits evaluation.

Betaherpesvirinae

Betaherpesvirinae includes cytomegalovirus (CMV), also called HHV-5, and herpes lymphotropic viruses HHV-6 and HHV-7. The CMV infection may involve lungs, liver, and spleen. The infection is often acquired through fecal–oral transmission and has affected the majority of persons by adulthood. It primarily targets epithelial cells, monocytes and lymphocytes. HHV-6, recognized since 1986, affects mostly during early childhood. Common symptoms include fever and macular lesions on skin (back and neck). About 50%–90% of cases are seropositive.

Table 17.1 Summary of Some Important Clinical Features of HSV-1 and HSV-2

Characteristics	HSV-1	HSV-2
Site of the primary infection:	Mouth, face, lips, and eyes	Genitals, thighs, and buttocks
CNS involvement:	Encephalitis	Meningitis
Mode of transmission:	Oral, bioaerosol	Sexual
Population at risk:	Pre-puberty	Post-puberty
Risk of blindness	Rare, but possible	Not uncommon

Gammaherpesvirinae

Subfamily Gammaherpesvirinae includes Epstein-Barr virus (EBV) and HHV-8. The EBV primarily targets epithelial and B cells while the HHV-8 mostly infects lymphocytes. HHV-8 is a Kaposi's sarcoma–related virus as well as related to primary effusion lymphoma and Castleman's disease. The EBV causes mononucleosis in children. The disease is usually asymptomatic, but cases involving upper respiratory tract infection, pharyngitis, tonsillitis, bronchitis, otitis media, and so on are not uncommon. In some cases, symptoms are accompanied by fever. About 50%–60% of Americans develop this disease at some point in life. The lab diagnosis is based on the demonstration of increased lymphocytes and specific antibodies. HSV, EBV, and CMV are also associated with the cases of hepatitis.

Poxviridae (Smallpox virus)

The family Poxviridae includes variola and vaccinia viruses, which are members the genus *Orthopoxvirus*. These are large, brick-shaped, complex viruses with a dumbbell-shaped core (Fig. 17.1) and measure $200–250 \times 300$ nm in size, hence an exception to the rule because they can be visualized through a bright field microscope. The variola virus causes smallpox in humans and the vaccinia virus is known as cowpox virus and used in the vaccine preparation. Like plague among the bacterial diseases, smallpox of all the viral diseases has left a distinct mark in the history of human civilization. Smallpox is perhaps the ugliest and the deadliest viral disease, responsible for hundredss of millions of deaths in Asia, especially in India, and the disfigurement of several times more men and woman, in the past millennia.

The infection is usually contracted via direct contact or through airborne particles from infected persons in close vicinity. The respiratory tract is the usual portal of entry. Initially, the virus colonizes the mucous membrane and then invades the lymph nodes. Incubation period ranges from 5 to 15 days. During the acute phase, the patient develops severe head and body aches accompanied by high fever. At this stage, the virus invades the capillary epithelium of the skin leading to the formation of papules, vesicles, and crusts. The fatality rate is approximately 50%. Those who survive are often left with ugly scars and disfigurement, and occasionally with the loss of vision in one or both eyes. The vaccination program initiated by WHO from the 1950s through the 1970s has succeeded in the eradication of smallpox from the face of the earth. However, large stockpiles of live smallpox virus are still present, officially at least under the direct control of the governments of Russia and United States, but unofficially perhaps in China and some other countries that formerly belonged to the Soviet Union. The potential for such live cultures falling in the hand of terrorists is high. If terrorists (agents of the state or individuals) manage to get hold of such cultures, they can easily use it as a weapon of mass destruction. Since a majority of today's population is not immunized, the consequences can be catastrophic.

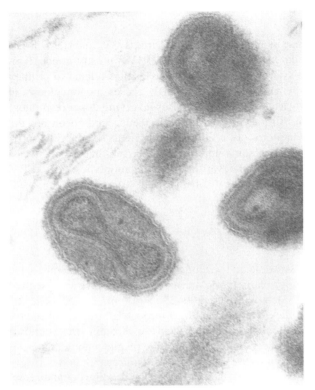

Figure 17.1.   Electron micrograph of smallpox virus, showing a dumbbell-shaped core (courtesy: CDC).

Hepadnaviridae (Hepatitis B Virus)

Hepatitis B virus (HBV) is an enveloped virus, has a rather complex structure, and multiplies in the nucleus. Three different sets of antigenic particles have been noted in the sera of persons infected with HBV. These include two spherical particles measuring 42 and 22 nm, and a tubular particle with a variable length. The HBV replicates in the hepatocytes. Hepatitis B is perhaps the most dreaded form of hepatitis. The incubation period may range from 8 to 24 weeks. Jaundice is seen in 25% cases and the mortality rate is about 1%–4%. By some estimates, currently 200 million persons are suffering from HBV infection in the world, more than 1 million in the United States. The common routes of infection include vaginal or anal sex, sharing needles during the IV drug usage, or intravenous infusion. During the course of infection, the envelope membrane is coated with an antigen called hepatitis B surface antigen (HBs antigen). Presence of HBs antigen in the body fluid is an indicator of hepatitis B infection. Use of a highly effective vaccine currently available is likely to dramatically reduce HBV infection during the coming years.

SINGLE-STRANDED DNA VIRUSES

Parvoviridae (Parvovirus)

Members of the family Parvoviridae are a unique group of viruses in the sense that their genome consists of only a single strand of DNA. These are naked, icosahedral viruses measuring approximately 23 nm in diameter, which multiply in the nucleus. Parvoviruses are best known for causing serious infections in dogs and cats. Only one member of the family Parvoviridae, tentatively called B19, is known to infect humans. The disease may be characterized by mild fever, headache, chills, and malaise, or a more serious complication in immunocompromised hosts. Occasionally, B19 infection in pregnant woman may lead to spontaneous abortion. The respiratory tract is believed to be the portal of entry and the incubation period may range from 4 to 14 days. Usually a lifelong immunity develops following the infection.

DOUBLE-STRANDED RNA VIRUSES

Reoviridae (Reo, Rota, and Orbi Viruses)

As is the case with ss-DNA viruses, there are not many ds-RNA viruses. The family Reoviridae includes Reovirus (Orthoreovirus), Rotavirus, and Orbivirus, which are all naked icosahedral viruses that measure 60–80 nm in diameter. Of the three, rotaviruses are important human pathogens that cause gastroenteritis in infants and young children. They account for nearly 50% of all cases of diarrhea in infants and children, often requiring hospitalization due to severe dehydration. It is believed that rotaviruses are responsible for more than a million deaths in poor countries. The infection begins with the adsorption of the viral particle to the columnar epithelium of the small intestine. Water absorption is reduced, resulting in watery diarrhea and loss of electrolytes. The symptoms are less severe in breastfed infants and children. Adults, if infected, seldom show any symptoms.

SINGLE-STRANDED RNA VIRUSES

Picornaviridae (Polio and Hepatitis A Virus)

Family Picornaviridae includes three important pathogens, Poliovirus, hepatitis A virus (HAV), and rhinovirus. These are naked, icosahedral, (+) sense RNA viruses that multiply in the cytoplasm. Hepatitis A virus can survive in water and can live for an extended period of time. The infection is acquired by ingestion of contaminated food and beverages. Initially the virus colonizes intestinal epithelium and eventually spreads to the liver. Symptoms lasting from 2 to 20 days include nausea, diarrhea, chills, and malaise. Jaundice may develop in some cases. The disease is associated with poor public hygiene. An effective vaccine, Havrix, prepared from dead HAV, is a useful prophylactic tool.

Like HAV, poliomyelitis or poliovirus can survive in water for an extended period, and the disease is acquired via ingestion of contaminated food and water. The incubation period is 6–20 days. The virus multiplies in the mucosa of throat and intestine, disseminates to tonsils and cervical lymph nodes, and then enters the blood stream. Symptoms if present are short-lived and include mild fever, headache, sore throat, and loss of appetite. In a small number of cases, the virus inters motor nerve cells of the spinal cord where it multiplies. The final outcome could be motor and muscle paralysis that may persist through the life of the patient. Three serotypes, P1, P2, and P3, have been noted. Extensive usage of Salk and Sabine vaccines during the past five decades has eradicated poliomyelitis from much of the industrialized world. An ambitious WHO program currently under way in the rest of the world has yielded encouraging results, but foci of poliomyelitis still persist in many remote areas of the world.

Rhinovirus is one of the several viruses that cause the common cold. The virus mostly infects nasal mucosa. Symptoms include nasal stuffiness, sneezing, and watery discharges from the nose. The disease is usually short-lived and self-healing. Because more than 100 serotypes exist, immunity is generally of a short duration. Reinfection by a different strain can occur, often within a brief span of time.

Caliciviridae (Hepatitis E Virus)

Members of the family Caliciviridae are naked, icosahedral, (+) sense RNA viruses that are assembled in the cytoplasm. A provisionally assigned member of this family is the hepatitis E virus (HEV), which is a food- and waterborne disease commonly noted in parts of Asia, Africa, and Latin America. Hepatitis E is usually a mild, self-limiting, and rarely fatal disease. The incubation period is 3–6 weeks. Jaundice is seen in about 80% infected adults, but it is somewhat rare (10%) in children. The disease is especially present in individuals consuming water with fecal contamination. The virus is quite stable and capable of surviving in food and water for a long time. ELISA and PCR are useful diagnostic tools. There is no specific therapy and no vaccine for immunization.

Togaviridae (Rubella Virus)

Members of the family Togaviridae are enveloped Icosahedral, (+) sense RNA viruses that multiply in the cytoplasm. An important member of this family is the rubella virus, which causes German measles in many parts of the world. The infection is highly contagious, and acquired through the inhalation of bioaerosols originating from an infected person. The incubation period is 12–23 days, and the usual symptoms include rashes (small red spots) accompanied by a mild fever. The symptoms typically last for 2–3 days and the disease is generally self-healing. However, rubella infection can have serious consequences when a pregnant woman is infected during the first trimester. Such infections can result in fetal death, premature delivery, or congenital defects involving the heart, eyes, or ears of the newborn. Widespread use of rubella vaccine (attenuated virus), usually as a part of the MMR (mump, measles and rubella) triad, has eradicated German measles in the United States and other industrialized countries. The disease still persists in societies with religious prohibitions against vaccines or countries with poor immunization programs.

Flaviviridae (Hepatitis C and G Viruses)

Members of the family Flaviviridae are enveloped, icosahedral, (+) sense RNA viruses that are assembled in the cytoplasm and measure 40–80 nm. This family includes causal agents of several viral diseases, such as yellow fever, dengue viruses, West Nile Fever (encephalitis) virus, St. Luis and Japanese encephalitis viruses, and hepatitis C (HCV), hepatitis G (HGV), and possibly hepatitis F (HFV) viruses. Of these, HCV, the causal agent of hepatitis C, has drawn greater attention in recent years. Globally, over a million cases are reported each year, nearly 25,000 new cases each year in the United States alone. The infection is acquired via transmission of the virus from an infected person through blood transfusion, organ transplantation, sexual intercourse or any other situation resulting in contact with HCV-containing body fluid. Placental transfer from infected mother to unborn fetus is also known to occur. Hepatitis C is a debilitating disease that can be fatal. It causes nearly 8,000 deaths each year in the United States, and also a major cause of liver transplantation. ELISA and PCR are important diagnostic tools. The clinical management is facilitated by ribavirin and interferon alpha. There are numerous side effects and response may often be incomplete. Protease inhibitors similar to those used in HIV management have recently been approved for the management of HCV. Unlike hepatitis A and B, there is no vaccine available for immunization against hepatitis C.

Another significant member of this family is the hepatitis G virus (HGV), which causes syncytial giant cell hepatitis and chronic liver inflammation. A virus suspected to cause fulminating post transfusion hepatitis and identified as hepatitis F virus (HFV) is believed to be closely related.

West Nile virus (WNV), also known West Nile encephalitis virus, is a zoonotic disease transmitted to humans from *Culex* mosquitoes. Certain birds, including crows and sparrows as well as a range of domestic and exotic birds, are often infected by the virus. *Culex* mosquitoes acquire it when they bite the birds and then transmit the virus to humans following a blood meal. Most infected individuals remain asymptomatic and clinical disease develops only in a small percentage of people. The disease was first noted in the West Nile district of Uganda in 1937 and subsequently encountered in several Middle Eastern, African, and Southwestern Asian countries. West Nile encephalitis was first noted in the United States in 1999 and by 2003, 9,800 cases and 264 deaths were noted. The U.S. Food and Drug Administration (FDA) has approved ELISA test for a rapid and reliable diagnosis of West Nile virus infection. There is neither a prophylactic vaccine nor treatment available to counter WNV infection.

Orthomyxoviridae (Influenza Virus)

The family Orthomyxoviridae includes the influenza viruses. These are enveloped, helical, and (−) sense RNA viruses that multiply in the cytoplasm. Influenza viruses are grouped into three categories, A, B, and C. The genome consists of 12,000 to 15,000 nucleotides. The segmented genome is divided into 8 physically separate molecules of the nucleic acid, which are packaged into a single virus particle. The capsid can be spherical, measuring 50–120 nm in diameter, or helical, measuring 200–300 nm in length and 20 nm in diameter. The envelope membrane surface contains hemagglutinin (HA) and neuraminidase (NA), which play critical roles in attachment and virulence. There are a total of 16 antigenic forms in HA (labeled as H1, H2, H3, etc.) and nine in the NA group (N1, N2, N3, etc.). These combine to produce different subtypes, such as H1N1, H2N3, and H5N1, which account for the seasonal variations in its antigenicity. Influenza viruses are widely distributed in nature and known to infect several mammalian and avian species. The disease causes high mortality and severe morbidity. Historically, influenza has caused several epidemics, often with a very high degree of mortality and morbidity. The 1918 epidemic was caused by H1N1, the 1957 epidemic by H2N2, and those in 1968 and 1977 were caused by H3N2 and H1N1 subtypes, respectively. Because the virus notoriously changes its antigenic profile, no successful vaccine that can provide long-term immunity has been developed. Therefore, influenza vaccines, often a cocktail of two or more subtypes, have to be formulated each year.

Paramyxoviridae (Mumps, Measles, and Respiratory Syncytial Viruses)

The family Paramyxoviridae includes Paramyxovirus (mumps virus), Morbillivirus (measles virus), and Pneumovirus (respiratory syncytial virus [RSV]). These are enveloped, helical viruses with a (−) sense RNA, which multiply in the cytoplasm. Mumps is an airborne infection that is acquired through the inhalation of infectious bioaerosol or direct contact with secretions from the infected subjects. The virus infects the nasopharynx and lymph nodes. Major symptoms include swelling of the salivary glands accompanied by mild fever. The disease occurs worldwide, but has been eradicated from a majority of the industrialized countries through the widespread use of attenuated mumps virus vaccine, which is usually a constituent of the MMR triad.

Measles virus, also known as Morbillivirus or rubeola virus, is the causal agent of the highly infectious disease measles. The portal of entry is usually the respiratory tract and occasionally the eye (conjunctiva). The incubation period is 10–21 days and the symptoms include nasal discharges, fever, headache, and conjunctivitis followed by skin eruptions occurring as pink macular lesions. Lesions may also develop in the oral cavity characterized by bright red spots with a bluish white center. There is no specific therapy for the measles. Like several other viral diseases, an attenuated vaccine, often as a component of the MMR triad, has been successfully used in the

United States. However, measles occurs all over the world and millions of children are affected each year in poor countries.

The respiratory syncytial virus (RSV) has variable size and shape, usually ranging from 120 to 300 nm in diameter. The capsid has two specific glycoproteins, namely G protein, responsible for the attachment, and F protein, which plays a role in the fusion of the cytoplasmic membranes of the infected cells. The fusion of infected cells results in the formation of syncytium which is actually a multinucleated mass of the fused cells. The disease results in infection of the lower respiratory tract, often necessitating hospitalization. Symptoms include high fever, cough, and rhinitis. The syncytia cause inflammation, alveolar thickening, and accumulation of fluid in the alveolar spaces. The infection is acquired through the inhalation of infectious bioaerosol or by direct contact with the clinical specimens. The disease occurs all over the world with considerable seasonal variations. There are no reliable vaccines.

Rhabdoviridae (Rabies Virus)

The most important member of the family Rhabdoviridae is the rabies virus, which is an enveloped, helical virus with a (−) sense RNA virus that multiplies in the cytoplasm. Rabies viruses measure 70–85 µm in diameter and 130–380 µm in length. Up to the beginning of the 20th century, rabies was quite common all over the world, but now rabies cases are mostly seen in less developed countries. The global mortality is believed to be in excess of 40,000 per year. Rabies is a zoonotic disease and several wild animals including wolves, foxes, coyotes, bats, and raccoons are often infected with this virus. Dogs, a frequent source of infection in humans, are believed to acquire the disease from one of these animals. The disease develops following a bite by the infected animals. The virus then enters the skeletal muscles, where it multiplies. Rabies virus is highly neurotropic. Depending on the amount of virus that enters the body, the symptoms develop in as little as 10 days or it may take 15–20 weeks before the symptoms fully develop. The early signs include anxiety, irritability, fatigue, and fever, followed by intense spasm of neck and chest muscles and eventually damage to the brain, leading to breathing failure. Multiple tools are available for the diagnosis of rabies. These include isolation of the virus in culture, ELISA, and direct immunofluorescence antibody test for postmortem examination of the brain tissue.

Effective vaccines, such as human diploid cell rabies vaccine or adsorbed rabies vaccine, are available. However, vaccines are effective only if administered soon after the animal bite. For prophylaxis, rabies vaccination is effective if given every 2 years.

Bunyaviridae (Hantaviruses)

Hantaviruses are perhaps the best known member of the family Bunyaviridae. These are enveloped, helical, (−) sense RNA viruses that multiply in the cytoplasm

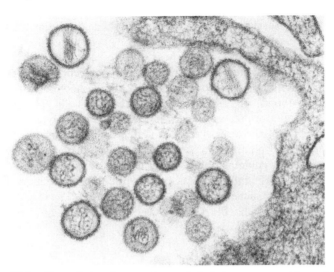

Figure 17.2.   Transmission electron micrograph depicting numerous virions of hantavirus in the cytoplasm (courtesy: CDC).

(Fig. 17.2). The disease, hantavirus pulmonary syndrome (HPS), is a recently recognized entity. It is characterized by hemorrhagic fever, respiratory distress, and renal syndrome. HPS is a strange mix of zoonotic and airborne infections. Certain rodents are believed to be the carrier of the virus. Infections are mostly acquired through the inhalation of aerosolized urine or fecal matter. The disease can be fatal if proper interventions are not made in a timely manner. Useful diagnostic tools include ELISA and reverse transcriptase-polymerase chain reaction. No specific therapy is available. Rodent control appears to be the only prophylactic tool.

Filoviridae (Ebola and Marburg Viruses)

The family Filoviridae includes Marburg and Ebola viruses. These are enveloped, helical, (−) sense RNA viruses that multiply in the cytoplasm (Fig. 17.3). They generally measure about 80 nm in diameter, and are nearly 1 μm long, which is quite an unusual length for a virus. These viruses are endemic in certain parts of Africa including Congo, Sudan, Zimbabwe, and Kenya. Wild monkeys are believed to be the natural reservoir, but in addition to monkey-to-human, human-to-human infections also occur. The disease is highly contagious. In the early phase of infection, flu-like symptoms are quickly followed by vomiting and diarrhea. In the advanced state, hemorrhage from several anatomical sites ensues. The disease is mostly fatal. Because of the fact that a significant number of persons in central Africa have antibodies to these viruses, at least theoretically, it is possible that subclinical and self-healing infections occur. However, there is no unequivocal evidence to support this

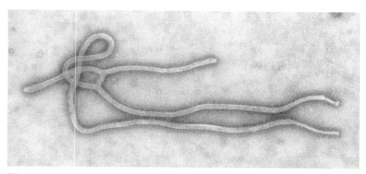

Figure 17.3.   Colorized transmission electron micrograph of Ebola virus (courtesy: CDC). See color insert.

notion. The laboratory diagnosis can be clinched by the use of ELISA, immunofluorescence, and reverse transcriptase-polymerase chain reaction. Cultures are possible but not advised unless Class 3 or Class 4 isolation facilities are available. There is no therapy for the Ebola virus infection. Interferon and antiserum have been tried without much success.

Retroviridae (HIV)

The family Retroviridae is generally divided into subfamilies Oncornavirinae and Lentivirinae. The subfamily Oncornavirinae includes human T-cell lymphotropic virus 1 (HTLV-I) and human T-cell lymphotropic virus 2 (HTLV-II). The subfamily Lentivirinae includes human immunodeficiency virus 1 (HIV-1) and human immunodeficiency virus 2 (HIV-2). These are icosahedral, enveloped, (+) sense RNA viruses that multiply in the cytoplasm. They usually measure 80–130 nm in diameter. The virion envelope contains spikes made of glycoproteins (gp4 and gp120), which play a crucial role in viral attachment to the host cells. The genome consists of a pair of RNA molecules and an enzyme called reverse transcriptase, which plays a critical role in its replication. The target cells are mostly T-helper cells and others that have CD4 protein on the surface of their plasma membrane. Macrophages, dendritic cells, and monocytes also have this binding site on their cell surface. Once the virus binds to the CD4 receptors, it enters the cell via a process called endocytosis. Following the uncoating (removal of the envelope), triggered by reverse transcriptase, the RNA is copied into a single stranded DNA molecule (reverse transcription). The RNA molecules are degraded and the DNA strand is duplicated to form a double-stranded DNA called "provirus." The provirus integrates with the host DNA and directs it to synthesize viral RNA which then directs synthesis of its own capsid protein. The assembly (insertion of the genome into capsid) takes place at the cytoplasmic membrane. The mature virus exits the cell through a process called budding. The host cell is eventually lysed and the newly released virion infects other cells having CD 4 proteins.

Of all the retroviruses, HIV-1 and HIV-2 have drawn greater attention for causing acquired immunodeficiency syndrome (AIDS) in humans. The majority of AIDS cases in the United States are attributed to HIV-1, and HIV-2 is mostly seen in association with AIDS in sub-Saharan Africa. The natural reservoirs cannot be identified with certainty, but chimpanzees are believed to be the ultimate source. Globally, approximately 40 million people are currently believed to be infected with HIV viruses, mostly in sub-Saharan Africa and East, Southeast, and South Asia. The incidence rate of HIV infection appears to be declining in recent years. The infection begins when the virus enters the human body through blood transfusion or fluid exchange via sexual intercourse, or needle sharing during the intravenous adminis-tration of recreational drugs. There is some controversy as to how HIV enters the bloodstream following intercourse with infected persons. Now we know that den-dritic cells bearing CD4 proteins are also present on the mucosal surfaces throughout the body. Most likely the infection begins at the mucous membrane itself from where it moves into the bloodstream. The early symptoms include headache, muscular rash, malaise, fever, and weight loss. Lymph node enlargement and oral candidiasis (thrush) soon follow. The assault on the immune system is severe and the patient begins contracting any of a wide range of infectious diseases, including infections by microorganisms hitherto considered harmless. Most symptomatic patients die within 2–3 years if left untreated. However, individuals infected with HIV may remain asymptomatic for 10 years or even longer.

For a provisional diagnosis, ELISA tests are generally used for the detection of HIV antibodies. The confirmation requires tests using western blot. The virus can also be isolated in culture, but this method is seldom used. As of now, there is no ability to completely eradicate AIDS. The current therapeutic approach is based on controlling opportunistic infection and malignancies, and reducing the viral popula-tion in the host. At present, a combination of three to four drugs, acting at different stages of viral replication, is used in order to reduce the viral load. These include a concurrent use of nucleoside reverse transcriptase inhibitors (zidovudine, Abacavir, tenofovir, etc.), nonnucleoside reverse transcriptase inhibitors (Nevirapine, efavi-renz, etc.), protease inhibitors (lopinavir, darunaviretc), and fusion inhibitors to prevent viral entry into the cells (fuzeon).

Another noteworthy RNA virus with uncertain affiliation is the hepatitis D virus (HDV). It is a ss-RNA virus and most likely the smallest of all RNA viruses. Also, it is unique in the sense that unlike others, it has a circular genome and depends on hepatitis B virus (HBV) for its capsid protein. This virus was first noted in 1977 as a delta agent responsible for the disease in a case of cytopathic hepatitis. It was later named hepatitis D virus. It replicates in liver cells infected with HBV, thus hepatitis D can be called a co-infection with hepatitis B. Hepatitis D is a severe disease with acute onset. The incubation period ranges from 3 to 13 weeks and onset of jaundice is variable. The chronic state (in 10%–15% of patients) may lead to hepatocellular carcinoma. Like HBV and HCV, HDV is also transmitted through contact with body fluids including IV infusion and sexual intercourse. The fatality rate is about 30%.

PRIONS

Prions represent one of the recently discovered pathogenic entities. In 1982, Stanley Prusiner first proposed the term "prions" for proteinaceous infectious particles as the causal agent of scrapie, a neurological disease in sheep. Mad-cow disease, also known as bovine spongiform encephalitis (BSE), is believed to be caused by prions. It appears that several human diseases, including kuru, Creutzfeldt-Jacob Disease (CJD), Gerstmann-Straussler-Scheinker syndrome, and fatal familial insomnia are also caused by prions. Prion-like infectious proteins are also believed to cause transmissible mink encephalitis (TME) in minks, feline spongiform encephalitis (FSE) in cats, and chronic wasting disease (CWD) in deer and elk. It may be just a matter of time until prion infections in humans are traced to these animals.

The disease is generally called transmissible spongiform encephalitis (TSE). The infection involves transformation of PrP^C, a normal host glycoprotein, into an unusually folded protein, named PrP^{Sc}. PrP^{sc} is a highly infectious protein. In humans, the gene Prnp that codes for the PrP^C protein is located on chromosome 20. PrP^C protein produced by the cells is expressed on the cell surface where it may react with PrP^{Sc} as a result of exposure to an infected animal or human and refold in the shape of the latter. The PrP^{Sc} thus produced enters the cell through endocytosis and accumulates in the cellular lysosomes. The mechanism of cell damage is poorly understood, but pieces of PrP^{Sc} molecule probably accumulate in the brain forming plaque-like large vacuoles; therefore, the term spongiform encephalitis. The symptoms are highly variable and diagnosis is mostly based on the postmortem histopathological examination of the brain.

The pathophysiology of TSE is not fully understood. Experimental studies have demonstrated that mice lacking Prnp genes (hence devoid of PrP^c protein) cannot be infected with PrP^{sc} proteins. However, animals with functional Prnp genes, hence having PrP^c protein, develop TSE when injected with PrP^{sc} or homogenized brain tissue from infected animals. Thus it appears that the normal protein PrP^c is required for infection with the PrP^{sc} protein. Prions are resistant to most antibiological physical and chemical agents. They can survive autoclaving at 135°C for 18 minutes or 600°C dry heat and ionizing radiation.

Bibliography and Suggested Reading

Abbas, A.K., Lichtman, A.H., and Pober, J.S. 2000. *Cellular and Molecular Immunology*, 4th ed. Philadelphia: W.B. Saunders.

Ahearn, D.G., Crow, S.A., Simmons, R.B., Price, D.L., Mishra, S.K., and Pierson, D.L. 1997. "Fungal colonization in a multi-story office building: Production of volatile organics." *Curr. Microbiol.* 35:305–308.

Ajello, L., and Hay, R.J. 1998. "Medical mycology." In Collier, L., Balows, A., and Sussman, M., editors, *Topley & Wilson's Microbiology and Microbial Infections*, vol. 4, 9th ed. New York: Hodder Arnold.

Aktories, K., and Barbieri, J.T. 2005. "Bacterial cytotoxins: Targeting eukaryotic switches." *Nat. Rev. Micribiol.* 3:397–410.

Albert, M.J. 1994. "Vibrio cholerae 0139 Bengal." *J. Clin. Microbiol.* 32:2345–2349.

Aldecoa-Otalora, E., Langdon, W.B., Cunningham, P., and Arno, M.J. 2009. "Unexpected presence of mycoplasma probes on human microarrays." *BioTechniques* 47:1013–1015.

Balows, A., Truper, H.G., Dworkin, M., Harder, W., and Schleifer, K.-H. (eds.). 1994. *The Prokaryotes*, 2nd ed. 4 vols. New York: Springer-Verlag.

Barry, A.L. 1999. "Antimicrobial resistance among clinical isolates of *Streptococcus pneumoniae* in North America." *Amer. J. Med.* 107:28S-35S.

Beaver, P.C., Jung, R.C., Cupp, E.W. 1984. *Clinical Parasitology*, 9th ed. Philadelphia: Lea & Febiger.

Belshe, R.B. (ed.). 1991. *Textbook of Human Virology*, 2nd ed. St. Louis, MO: Mosby.

Bengis, R.G., Leighton, F. A., Fischer, J.R., Artois, M., Morner, T., and Tate, C.M. 2004. "The role of wildlife in emerging and re-emerging zoonoses." *Rev. Sc. Tech.* 23:497–511.

Bergey's Manual of Systematic Bacteriology, 2nd ed. 2001–2011. 5 vols. New York: Springer-Verlag.

Bergey's Manual of Determinative Bacteriology, 9th ed. 1994. Philadelphia: Williams & Wilkins.

Black, C. 1997. "Current methods of laboratory diagnosis of *Chlamydia trachomatis* infections." *Clin. Microbiol. Rev.* 10:160–184.

Boman, J., Gaydos, C., and Quinn, T. 1999. "Minireview: Molecular diagnosis of *Chlamydia pneumoniae* infections." *J. Clin. Microbiol.* 37:3791–3799.

Braude, A.I., Davis, C.E., and Fierer, J. (ed.). 1981. *International Textbook of Medicine*, vol. 2: *Infectious Diseases and Medical Microbiology*. Philadelphia: W.B. Saunders.

Buchanan, J. 1897. *An Encyclopedia of the Practice of Medicine Based on Bacteriology*, 7th ed. New York: R.R. Russell.

Burge, H.A., Pierson, D.L., Groves, T.O., and Mishra, S.K. 2000. "Dynamics of airborne fungal populations in a large office building." *Curr. Microbiol.* 40:10–16.

Cherry, J., and Robbins, J. 1999. "Pertussis in adults: Epidemiology, signs and symptoms and implications for vaccination." *Clin. Infect. Dis.* 28 (suppl. 2).

Cohen, J., Powderly, W., and Opal, S.M. 2010. *Infectious Diseases*, 3rd ed. St. Louis, MO: Mosby.

Cunningham, M. 2000. "Pathogenesis of group A streptococcal infections." *Clin. Microbiol. Rev.* 13:470–511.

Curtis, R. 2002. "Bacterial infectious disease control by vaccine development." *J. Clin. Invest.* 110:1061–1066.

Dalton, H.P., and Nottebart, H.C. 1986. *Interpretive Medical Microbiology*. New York: Churchill Livingstone.

A Concise Manual of Pathogenic Microbiology, First Edition. Saroj K. Mishra and Dipti Agrawal.
© 2013 Wiley-Blackwell. Published 2013 by John Wiley & Sons, Inc.

Dumler, J.S. 1996. "Laboratory diagnosis of human rickettsial and ehrlichial infections." *Clin. Microbiol. Newsletter.* 18:57–61.

Eiden, M., Buschmann, A., Kupfer, L., and Groschup, M.H. 2006. "Synthetic prions." *J. Vet. Med.* 53:251–256.

Emmons, C.W., Binford, C.H., Utz, J.P., and Kwon-Chung, K.J. 1977. *Medical Mycology*, 3rd ed. Philadelphia: Lea & Febiger.

Espy, M., Uhl, J., Sloan, L., Buckwalter, S., Jones, M., et al. 2006. "Real time PCR in clinical microbiology: Application for routine laboratory testing." *Clin. Microbiol. Rev.* 19:165–256.

Facklam, R., and Elliot, J.A. 1995. "Identification, classification, and clinical relevance of catalase-negative, Gram-positive cocci, excluding streptococci and enterococci." *Clin. Microbiol. Rev.* 8:479–495.

Fang, A., Pierson, D.L., Mishra, S.K., Koenig, D.W., and Demain, A.L. 1997. "Secondary metabolism in simulated microgravity: Beta-lactam production by *Streptomyces clavuligerus*." *J. Industrial Microbiol.* 18:22–25.

Feldman, H.A. 1986. "The meningococcus: A twenty year perspective." *Rev. Infect. Dis.* 8:288–294.

Forbes, B.A., Sahm, D.F., and Weissfeld, A.S. 2007. *Bailey & Scott's Diagnostic Microbiology*, 12th ed. St. Louis, MO: Mosby.

Garcia, L.S. (ed.). 2001. *Diagnostic Medical Parasitology*, 4th ed. Washington, D.C.: American Society for Microbiology.

George, M.J. 1995. "Clinical significance and characterization of *Corynebacterium* species." *Clin. Microbiol. Newsletter* 17:177–180.

Groisman, E., and Casadesus, J. 2005. "The origin and evolution of human pathogens." *Mol. Microbiol.* 56:1–7.

Gutierrez, Y. 1990. *Diagnostic Pathology of Parasitic Infections with Clinical Correlations*. Philadelphia: Lea & Febiger.

Hamon, M., Bierne, H., and Cossart, P. 2006. "*Listeria monocytogenes*: A multifaceted model." *Nat. Rev. Microbiol.* 4:423–434.

Heikens, E., Fleer, A., Paauw, A., Florijin, A., and Fluit, A.C. 2005. "Comparison of genotypic and phenotypic methods for species-level identification of clinical isolates of coagulase negative staphylococci." *J. Clin. Microbiol.* 43:2286–2290.

Hook, E.W., and Holmes, K.K. 1985. "Gonococcal infections." *Ann. Intern. Med.* 102:229–243.

Howard, B.J., Klaas, J., Rubin, S.J., Weissfeld, A.S., Tilton, R.C. 1994. *Clinical and Pathogenic Microbiology*. St. Louis, MO: Mosby.

Huang, J., Putnam, A., Werner, G.M. Mishra, S.K., and Whitenack, C. 1989. "Herbicidal metabolites from a soil-dwelling fungus (*Scopulariopsis brumptii*)." *Weed Science* 37:123–128.

Huebner, R.E., and Castro, K.G. 1995. "The changing face of tuberculosis." *Amer. Rev. Med.* 46: 47–55.

Hutchison, C.A., and Montague, M.G. 2002. "Mycoplasmas and the minimal genome concept." In Razin, S., and Herrmann, R., editors, *Molecular Biology and Pathogenicity of Mycoplasmas*. New York: Kluwer Academic/Plenum.

Isenberg, H.D. 1992. *Clinical Microbiology Procedures Handbook*, 2nd ed. 2 vols. Washington, DC: American Society for Microbiology.

Jahrling, P.B., Fritz, E.A., and Hensley, L.E. 2005. "Counter measures to the bioterrorist threat of smallpox." *Curr. Mol. Med.* 5:815–826.

Kafetzis, D.A., Skevaki, C.L., Skouteri, V., Gavrili, S., Peppa, K., Kostalos, C., Petrochilou, V., and Michalas, S. 2004. "Maternal genital colonization with Ureaplasma urealyticum promotes preterm delivery: Association of the respiratory colonization of premature infants with chronic lung disease and increased mortality." *Clin. Infect. Dis.* 39:1113–1122.

Kelly, C., and LaMont, J.T. 1998. "*Clostridium difficile* infection." *Ann. Rev. Med.* 49:375–390.

Kerr, J., and Mathews, R. 2000. "Bordetella pertussis infection: Pathogenesis, diagnosis, management, and the role of protective immunity." *Eur. J. Clin. Microbiol. Infect. Dis.* 19:77–88.

Klingler, J. M., Stowe, R.P., Obenhuber, D.C., Groves, T.O., Mishra, S.K., and Pierson, D.L. 1992. "Evaluation of biological automated microbial identification system." *Appl. Environ. Microbiol.* 58:2089–2092.

Knapp, J.S. 1999. "Antimicrobial resistance in *Neisseria gonorrhoeae* in the United States." *Clin. Microbiol. Newsletter.* 21:1–7.

La Scola, B., and Raoult, D. 1997. "Minireview: Laboratory diagnosis of rickettsioses: Current approach to diagnosis of old and new rickettsial diseases." *J. Clin. Microbiol.* 35:2715–2727.

Larsen S., Steiner, B., and Rudolph, A. 1995. "Laboratory diagnosis and interpretation of tests for syphilis." *Clin. Microbiol. Rev.* 8:1–21.

Lederberg, J. (ed.). 1992. *Encyclopedia of Microbiology.* 4 vols. San Diego, CA: Academic Press.

Lennette, E.H. (ed.). 1999. *Laboratory Diagnosis of Viral Infections,* 3rd ed. New York: Marcel Dekker.

Lo, S. 1995. "New understandings of mycoplasmal infections and diseases." *Clin. Microbiol. Newsletter* 17:169–173.

Mahon, C.R., and Manuselis, G. 2010. *Textbook of Diagnostic Microbiology,* 4th ed. Philadelphia: W.B. Saunders.

Mandell, G., Bennett, J.E., and Dolin, R. 2010. *Principles and Practice of Infectious Diseases,* 7th ed. New York: Churchill Livingstone.

Manz, R.A., Hauser, A.E., Hiepe, H., and Radbruch, A. 2005. "Maintenance of serum antibody levels." *Annual Rev. Immunol.* 23:367–386.

Markell, E.K., John, D.T., Krotoki, W.A. (eds.). 1999. *Markell and Voge's Medical Parasitology,* 8th ed. Philadelphia: W.B. Saunders.

McGinnis, M. R., Molina, T.C., Pierson, D.L., and Mishra, S.K. 1996. "Evaluation of the Biolog Microstation System for yeast identification." *J. Med. Vet. Mycol.* 34:349–352.

Mehta, S.K., Stevens, D.A., Mishra, S.K., Firoze, F., and Pierson, D.L. 1999. "Distribution of *Candida albicans* genotypes among family members." *Diag. Microbiol. Infect. Dis.* 34:19–25.

Mishra, S.K. 1984. "Antigenic profile of some typical and septate phialides-strains of *Aspergillus fumigatus.*" *J. Med. Vet. Mycol.* 22:91–100.

Mishra, S.K. 1996. "Effect of space flight on humoral immune response to *Candida albicans.*" *JSC Reasearch and Technology Annual Report,* NASA Technical Memorandum 102795, 0. 57–59, NASA, Washington, DC.

Mishra, S.K., and Gordon, R.E. 1981. "*Nocardia* and *Streptomyces.*" In Braude, A.I., Davis, C.E., and Fierer, J., editors, *International Textbook of Medicine,* vol. 2: *Medical Microbiology and Infectious Diseases.* Philadelphia: W.B. Saunders.

Mishra, S.K., and Randhawa, H.S. 1969. "Application of paraffin bait technique to the isolation of *Nocardia asteroides* from clinical specimens." *Appl. Microbiol.* 18:686.

Mishra, S.K., Gordon, R.E., and Barnett, D.A. 1980. "Identification of nocardiae and streptomycetes of Medical Importance." *J. Clin. Microbiol.* 11:728–736.

Mishra, S.K., Staib, F., Folkens, U., and Fromtling, R.A. 1981. "Serotypes of *Cryptococcus neoformans* strains isolated in Germany." *J. Clin. Microbiol.* 14:106–107.

Mishra, S.K., Staib, F., Rajendran, C., and Folkens, U. 1982. "Serodiagnositc value of culture filtrate antigens from aspergilli with septate phialides." *Sabouraudia* 20:63–74.

Mishra, S.K., Keller, J.E., Miller, J.R., Heisey, R.M., Nair, M.G., and Putnam, A.R. 1987a. "Insecticidal and nematicidal properties of microbial metabolites." *J. Industrial Microbiol.* 2:267–276.

Mishra, S.K., Taft, W.H., Putnam, A.R., and Ries, S.K. 1987b. "Plant growth regulatory metabolites from novel actinomycetes." *J. Plant Growth Regul.* 6:75–84.

Mishra, S.K., Whitenack, C.J., and Putnam, A.R. 1988. "Herbicidal properties of secondary metabolites from several genera of soil microorganisms." *Weed Science* 36:122–126.

Mishra, S.K., Ajello, L., Ahearn, D., Burge, H., Kurup, V.P., Pierson, D.L., Price, R., Sandhu, R.S., Shelton, B., Simmons, R.B., and Switzer, K.F. 1992. "Environmental mycology and its importance to human health." *J. Med. Vet. Mycol.* 30 (Suppl. 1):287–305.

Mishra, S.K., Segal, E. Gunter, E., Kurup, V.P., Mishra, J., Murli, P.S., Pierson, D.L., Sandovsky-Losica, H., and Stevens, D.A. 1994. "Stress, immunity and mycotic diseases." *J. Med. Vet. Mycol.* 32 (Suppl. 1):379–406.

Nair, M.G., Putnam, A.R., Mishra, S.K., Mulks, M.H., Taft, W.H., Keller, J. E., and Miller, J.R. 1989. "Faeriefungin: A new broad-spectrum antibiotic from *Streptomyces griseus* var. *autotrophicus.*" *J. Nat. Products* 52:797–809.

Nair, M.G., Mishra, S.K., and Putnam, A.R. 1992. "Antifungal anthracycline antibiotics, spartanamicins A and B from *Micromonospora* species." *J. Antibiotics* 45:1738–1745.

Nataro, J.P., and Kaper, J.B. 1998. "Diarrheagenic *Escherichia coli*." *Clin. Microbiol. Rev.* 11:142–201.

Oshima, K., and Nishida, H. 2007. "Phylogenetic relationships among mycoplasmas based on the whole genomic information." *J. Mol. Evol.* 65:249–58.

Pierson, D.L., Mehta, S.K., Magee, B.B., and Mishra, S.K. 1995. "Person-to-person transfer of *Candida albicans* in the spacecraft environment." *J. Med. Vet. Mycol.* 33:145–150.

Pierson, D.L., M. Chidambaram, M., Heath, J.D., Mallary, L., Mishra, S.K., Sharma, B., and Weinstock, G.M. 1996. "Epidemiology of *Staphylococcus aureus* during space flight." *FEMS Immunol. Med. Microbiol.* 16:273–281.

Plotkin, S.A., and Orenstein, W.A. 1999. *Vaccines*, 3rd ed. Philadelphia: W.B. Saunders.

Podschun, R., and Ulmann, U. 1998. "*Klebsiella* spp. as nosocomial pathogens: Epidemiology, taxonomy, typing methods, and pathogenicity factors." *Clin. Microbiol. Rev.* 11:589–603.

Richman, D.D., Whitley, R.J., and Hayden, F.G. 1997. *Clinical Virology*. New York: Churchill Livingstone.

Rippon, J.W. 1988. *Medical Mycology: The Pathogenic Fungi and the Pathogenic Actinomycetes*, 3rd ed. Philadelphia: W.B. Saunders.

Rupp, M.E., and Archer, G.L. 1994. "Coagulase-negative staphylococci: Pathogens associated with medical progress." *Clin. Infect. Dis.* 19:231–245.

Russel, D.G. 2007. "Who puts the tubercle in tuberculosis?" *Nat. Rev. Microbio.* 5:39–47.

Schlech, W. 2000. "Foodborne listeriosis." *Clin. Infect. Dis.* 31:770–775.

Soll, A. 1996. "Medical treatment of peptic ulcer disease: Practice guidelines." *J.A.M.A.* 275:622–629.

Staib, F., and Mishra, S.K. 1982. "Extracellularly proteolysing *Aspergillus fumigatus* and its interaction with specific immune serum in the serum albumin agar." *Zbl. Bakt. Hyg., I. Abt. Orig. A* 253: 272–278.

Staib, F., Mishra, S.K., Grosse, G., and Abel, T. 1977. "Pathogenesis and therapy of cryptococcosis in animal experiments." In Beemer, A.M., et al., editors, *Host–Parasite Relationships in Systemic Mycoses, Contributions to Microbiology and Immunology*. Vol. 3. Basel, Switzerland: Karger, pp. 43–59.

Staib, F., Rajendran, C., Mishra, S.K., Voigt, R., Lindlar, F., and Hartman, C. 1983. "An atypical *Aspergillus flavus* from a case of bronchopulmonary aspergilloma." *Zbl. Bakt. Hyg., I. Abt. Orig. A* 255: 361–367.

Tarkowski, T.A., Koumans, H.E., Sawyer, M., Pierce, A., Black, C.M., Papp, J.R., Markowitz, L., and Unger, R.E. 2004. "Epidemiology of human papillomavirus infection and abnormal cytologic test results in an urban adolescent population." *J. Infect. Dis.* 189:46–50.

Versalovic, J., Carrol, K.C., Funke, G., Jorgensen, J.H., Landry, M.L., and Warneck, D.W. 2011. *Manual of Clinical Microbiology*, 10th ed. 2 vols. Washington, DC: American Society for Microbiology.

Waldvogel, F.A. 1999. "New resistance in *Staphylococcus aureus*." *N. Eng. J. Med.* 340:556–557.

Index

A Concise Manual of Pathogenic Microbiology, First Edition. Saroj K. Mishra and Dipti Agrawal.
© 2013 Wiley-Blackwell. Published 2013 by John Wiley & Sons, Inc.